Connecting The Dots To Future Electric Power

Edward J. Bair

Bloomington, IN

authorHOUSE®

Milton Keynes, UK

AuthorHouse™
1663 Liberty Drive, Suite 200
Bloomington, IN 47403
www.authorhouse.com
Phone: 1-800-839-8640

AuthorHouse™ UK Ltd.
500 Avebury Boulevard
Central Milton Keynes, MK9 2BE
www.authorhouse.co.uk
Phone: 08001974150

First published by AuthorHouse 4/23/2007

ISBN: 978-1-4259-9586-7 (sc)
ISBN: 978-1-4259-9585-0 (hc)

Library of Congress Control Number: 2007901012

Printed in the United States of America
Bloomington, Indiana

This book is printed on acid-free paper.

Acknowledgements

My wife Dorothy served nobly as text editor, content editor, and advisor. Her fine arts education qualifies her as an outside arbiter of what is readable by non-specialists. More importantly she made the effort to prove that the ideas are understandable to non-specialists even though they are not always easy reading.

My friend Dr Donald A. Ramsay of the Canadian National Research Council read an entire draft of the manuscript and offered criticism that led to major revisions.

Charles Matsen of the Indiana University Physical Plant introduced me to the world of independent service operators.

Professor Victor Viola, Dr. Henry Kramer, and Dr. Alan Waltar helped resolve my misconceptions about breeder reactors.

Mr. Robin Parker and Dr. Peter Langhoff of Solar Reactor Technologies, Inc. and Mr. David Rib of the Kramer Junction Company introduced me to the U.S. solar power capabilities.

Ralph Nansen, the space power commissioner from Boeing, gave helpful criticism of early versions of Chapter 11.

Many colleagues at Indiana University read selected parts of the manuscript and made useful comments. They include Professors Robert Bent, Bennett Brabson, Marvin Carmack, David Clemmer, Gary Hieftje, Kenneth Mantei, Carlos Miller, Lloyd Orr, Charles Parmenter, Scott Robeson, Nelson Schaffer, Philip Stevens, and David Winkle.

Many members of the staff of the Indiana University Library system gave me a new appreciation for what good librarians do.

Foreword

People who rely on technology to solve all future problems need to understand the limits of nature. The misunderstanding about future electrical power is mostly a fundamental human difficulty in comprehending large quantities. Mathematics is about the logic of numbers. The logic is independent of whether the numbers have meaning. The numbers range from infinitely small to infinitely large.

In the real world the same numbers and logic represent quantities. The concept of an infinite quantity is beyond human comprehension. Electric power involves quantities that are finite but large enough to deserve some forethought about the practical aspects of very large quantities.

Science and technology attach symbols to numbers to identify the units of measurement. These symbols obey the same rules of algebra as the numbers themselves. They imply an answer to the question, "Quantity of what?" Flow of electric power is measured in *watts*, W. Energy is power over a given time (power x time); *watt-hours* or Wh. The same units can also represent a quantity of heat. Converting heat to electric power typically requires 3 times more heat than the power it produces. A subscript on the dimension symbol, W_h or W_e, can make the distinction.

Quantities are inherently statistical and have a range of uncertainty. The quantities associated with future electric power are seldom precise but have a range of over 14 powers of ten. Three examples show that the multiplier of the dimensions not only specifies the magnitude but changes the concept.

One kilowatt (1000 watts or 1 kW) is the electric power per person developed nations consume continuously. It is double the power an average person can exert with maximum effort on a treadmill, pulling a plow, or generating electric power. World population limits the arable land to about an acre per person. The abundance of nature no longer enables people to prosper on a small fraction of a kW of muscle power. Electric power gives people the productivity to live on a decreasing share of resources. This makes electric power demand irreversible.

One gigawatt or GW is a million kW. It is the output of a power plant that supplies a large city. The 1 kW per day of electric power a person consumes can be supplied by 4 kg of coal, a quantity a child can carry with one hand. One GW of power per day takes 2 freight trains or 80 gondola cars, of coal. One kW and one GW are not only different numbers. They are different concepts.

In groups larger than about 100 units lose individual identity. The 500 GW of power North America consumes is different from the one GW a power plant produces or the one kW an individual consumes. The world will consume 5000 GW of power if the average consumption reaches as much as a half kW per person.

One petawatt or PW is a million GW. This is beyond the range of electric power. It is in the range of power nature provides. The Earth receives 173 PW of radiant power from the Sun. Before thermal radiation or the albedo return it to space it creates rainfall, rivers, lakes, plant and animal growth, weather, climate, and much of what makes the Earth different from the Moon. On Earth it is a practical value for infinite power. It is 100,000 times the electrical power we now consume. Only a small fraction of the available power is accessible for electricity.

Present civilization system is scavenging scraps of energy left from the past history. Each year the need for energy increases more than the population. Each year the quality of the available fuel resources decreases. We now use the best million years of energy nature took 500 million years to store. The future will require even more, perhaps much more.

The only choice is to rely on the diminishing quality of energy resources that have been stored by nature until ways can be found to capture about 5000 GW of the 121 PW that reaches the Earth. The sources of renewable energy are relatively obvious. The arithmetic by which they produce significant power is not.

Contents

Chapter 1
Economic Basis of Electric Power

People have built great civilizations for 40 centuries. Electric power did not appear until a century ago. Yet dependence on power already seems irreversible. Most people take it for granted that technology will provide power in the future. This is not as certain as it may seem. Human history clarifies why electric power is necessary and worth great effort.

A compressed sense of history is partly a normal effect of time and partly a real increase in the pace of truly historic events. Knowledge growth compounds an already distorted importance of current history. The long view shows the distinction.

The Earth formed 4.5 billion years ago. An environment habitable to biological organisms did not start to form on land until a half billion years ago. Geologists divide the formation of a habitable environment into increments of 20-80 million years. Fossil remains of plants and animals we now use for energy accumulated over this entire period. They also provide the historical evidence.

Human history is recent, probably the past million years[1, 2]. Ancestors of humans that now exist are traceable to a migration out of Northern Africa 100 millennia ago. Over the next 40 millennia that migration scattered over most of the Earth[3, 4]. Evidence indicates that these humans were clever and well adapted to using tools. To survive harsh climates they invented clothing. Electric power may be an adopted necessity like clothing.

Human ability to transmit experience to future generations in writing evolved over time. Cave art from 50 millennia ago in Europe is early evidence[5]. Written languages began independently in different societies. Symbolic language is an inherent defining human trait.

Cultural evolution goes beyond biological evolution. At birth babies understand little about the world they live in. They must learn. The more they learn the more they can teach the next generation. Written knowledge is both cumulative and combinative. New knowledge in one area can multiply the knowledge of many areas. Knowledge grows geometrically.

Human Population

Fig. 1.1 shows the history of human population. It is an aspect of modern humans like the elephant in the living room people prefer not to notice. The late Biology Professor Dean Fraser demonstrated by placing a single bacterium in a Petri dish of nutrient[6]. Overnight the population of bacteria would surpass the current world human population. After a day or so the nutrient would be gone and the bacteria would be dead. It is a stark illustration of the consequences of unsustainable growth in an environment with limited resources.

The doubling time is one way to express the rate of population growth. The doubling rate of bacteria under favorable circumstances is constant, about 20 minutes. The doubling rate of humans is not. Until the start of agriculture about ten millennia ago it took an average of 30-50 centuries for world population to double. From then up to 2000 years ago the population doubled every 13-19 centuries on average. From 1 to 1650 AD it took an average of 2-4 centuries to double. From 1650 to the present it has doubled every 50 years[7]. After nearly 100 millennia of almost negligible growth the rate switched to a hundred-fold increase.

Recent population growth makes a demographic transition to more sustainable growth inevitable. This can occur in only two ways. The birth rate can decrease or the rate of premature death can increase. Both alternatives are probably already in progress. In developed nations the birth rate is decreasing. There is a plausible reason for this to be so. A major cause of the population increase is increased life expectancy. In underdeveloped nations the rate of premature death due to disease, starvation, and genocide is now beginning to cancel the gains in life expectancy.

A catastrophic extinction of the human population is possible. The dinosaurs and wooly mammoths are precedents. A plan to prevent this may not be something humans can accomplish. Nevertheless population is a necessary consideration in any long range plan in any human policy objectives.

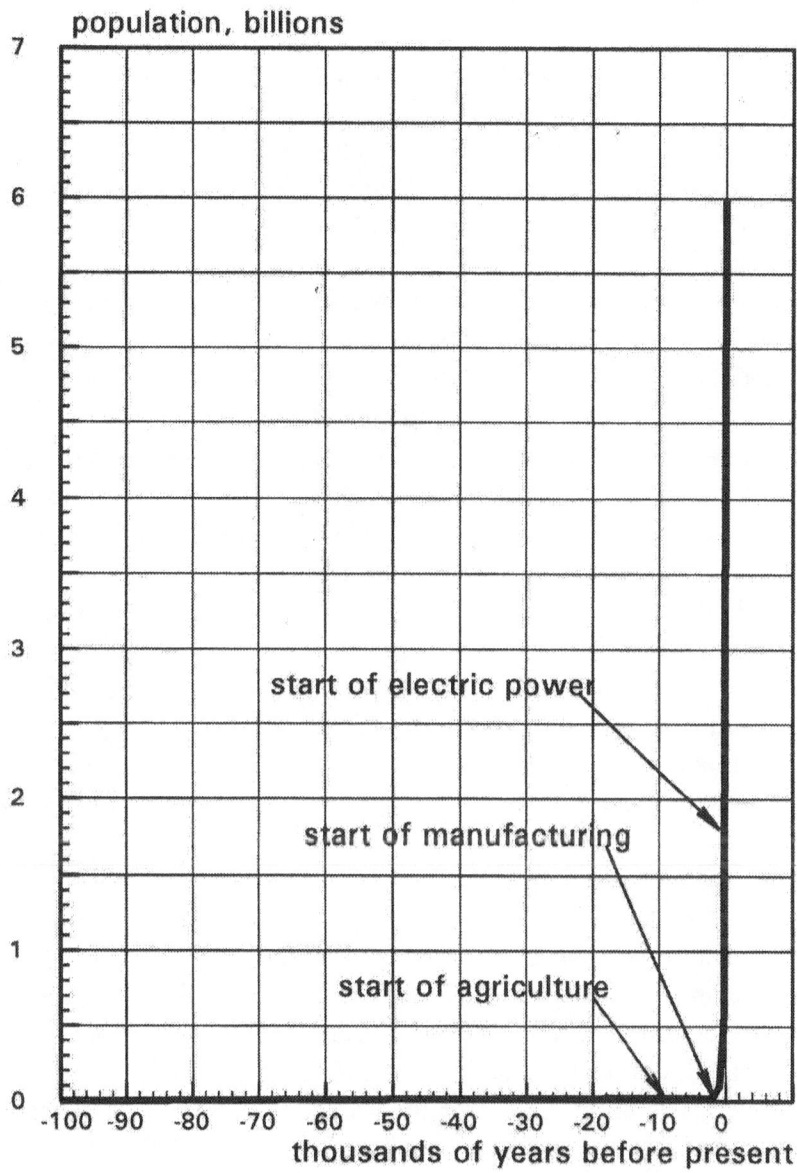

population, billions

start of electric power

start of manufacturing

start of agriculture

thousands of years before present

Fig. 1.1 History of human population growth.

How People Earn Income

Agriculture was the first major departure from passive dependence on nature. It began independently in Iraq, China, and Latin America about 10 or 11 millennia ago. Relatively few places had the required indigenous grains and animals capable of domestication for food or work[8]. Once established it had great advantages. People could have permanent homes and other wealth they did not have to carry with them. They had time for inventions. Greater productivity and more specialized ways to earn a living became a basis for centers of organization and wealth. These centers arose more or less independently throughout all temperate climates.

The Earth's surface is 30% land. Climate, terrain, and soils limit the arable land to about 13% of the total[9, 10]. An additional fraction is suitable for grazing or forestry. The number of people per acre of arable land is a measure of *population density*.

Where geography did not prevent it agriculture spread to adjacent lands at the same latitude. Food crops adapted to small differences in climate. They spread from the fertile triangle of Iraq west to Europe and east to the Malay Peninsula.

India has a favorable climate and a large fraction of its land is arable. Its population density is moderate in spite of its size. Russia has a low population density due to harsh climate and short growing season. China has a long agricultural history. Only a small fraction of its land is arable. It supports a very high population density by intensive land use. The climate allows several crops per year. North America and Australia have low population density due to relatively recent population growth. The United States has about the same area as China but double the fraction of arable land.

Africa is unfavorable for the spread of agriculture. Most of its native plants and animals are not capable of domestication. Its north-south orientation together with the Sahara desert blocks the spread into sub Sahara Africa. As a result much of Africa retained many aspects of a hunter-gatherer society well into the 20th century.

Industrialization was the second major human departure from passive dependence on nature. It began in Europe with the invention of heat engines three centuries ago. Factories that could support the cost of a steam engine applied power using belts, pulleys, and gears to achieve large economies of scale. The factories centralized the manufacturing work force.

Electric power appeared a little over a century ago. It is simpler to transmit power by wires than belts and pulleys. Inventions that replace workers by machines are easier to make. A power grid gives universal access at any location. Productivity increased throughout the economic system. This caused major changes that affected the labor force. A few people required new skills and more knowledge. Others became obsolete. World demographics reflect the disruptions that spread from their origins in Northern Europe.

Table 1.1 compares population growth, density and fraction in agriculture in the major world geographic regions[11]. Much data for the Russian Federation may be an anomaly of the Cold War.

Table 1.1 Population growth, density, and employment in agriculture

	% pop growth	people/acre	% ag pop
Sub-Sahara Africa	2.33	1.83	68
Middle East	1.75	1.92	35
Indian Subcontinent	1.63	2.63	64
Latin America	1.27	1.73	21
East Asia	1.03	5.03	45
Australia / N.Z.	1.01	.20	6
U.S. / Canada	0.91	.69	3
Europe	0.19	1.64	9
Russian Federation	-0.08	.51	23

Data Source: Central Intelligence Agency Factbook 2004

The residual population density appears to reflect past history more than occupational change. Agriculture productivity increased like all other sectors. The fraction of the work force in agriculture is smaller in the regions most affected by industrialization. In an economy based on knowledge education takes a long time at high cost. The advantage of a large family decreases. This is consistent with lower population growth.

The gross domestic product or GDP is the sum of the value of the goods and services the people of a nation produce. It is now compiled by every nation[12]. The total GDP is a measure of economic power. The GDP per person is a measure of the average individual productivity or income[13]. There are two ways to convert the GDP of other nations into U.S. dollars. The *currency exchange* value uses the currency exchange rate. The *purchase price parity* or *ppp* value uses the sum of the cost each item would have in U.S. dollars in that nation. This gives a more realistic relative standard of living in underdeveloped countries. The two values converge as trade establishes a global market price.

Fig. 1.2 shows the trend of occupational productivity. The bar graph of 150 largest nations decreases in order of GDP/person. Each bar has three widths corresponding to the contribution by people in agriculture, manufacturing, and other services.

The wide bottom segment is the fraction of the GDP/person due to the agriculture work force. The dollar amount varies by only about a factor-of-2 among different nations. The average is $575/person. The average annual income of farmers is $575/y divided by the agriculture fraction of the population in that nation. If 50% of the people earn their income in agriculture this is $1150/y. In a nation with a small percentage of farmers this can average $30,000.

The center portion of each bar is the fraction of the population with income from manufacturing. In a typical developed nation the agriculture work force contributes about 5% and the manufacturing work force about 25% of the GDP. In manufacturing as in agriculture increased productivity decreases the fraction of the population employed. The basic need for agricultural and manufactured products continues. Increased productivity in agriculture and manufacturing require increased support from other services.

The top section of each bar is the contribution by people in the service sector. This includes everyone who earns income in ways other than agriculture and manufacturing. In wealthy nations this is more than 70% of the population. Human productivity first switched from a hunter gatherer mode to agriculture. Fig. 1.2 shows the change toward greater diversification.

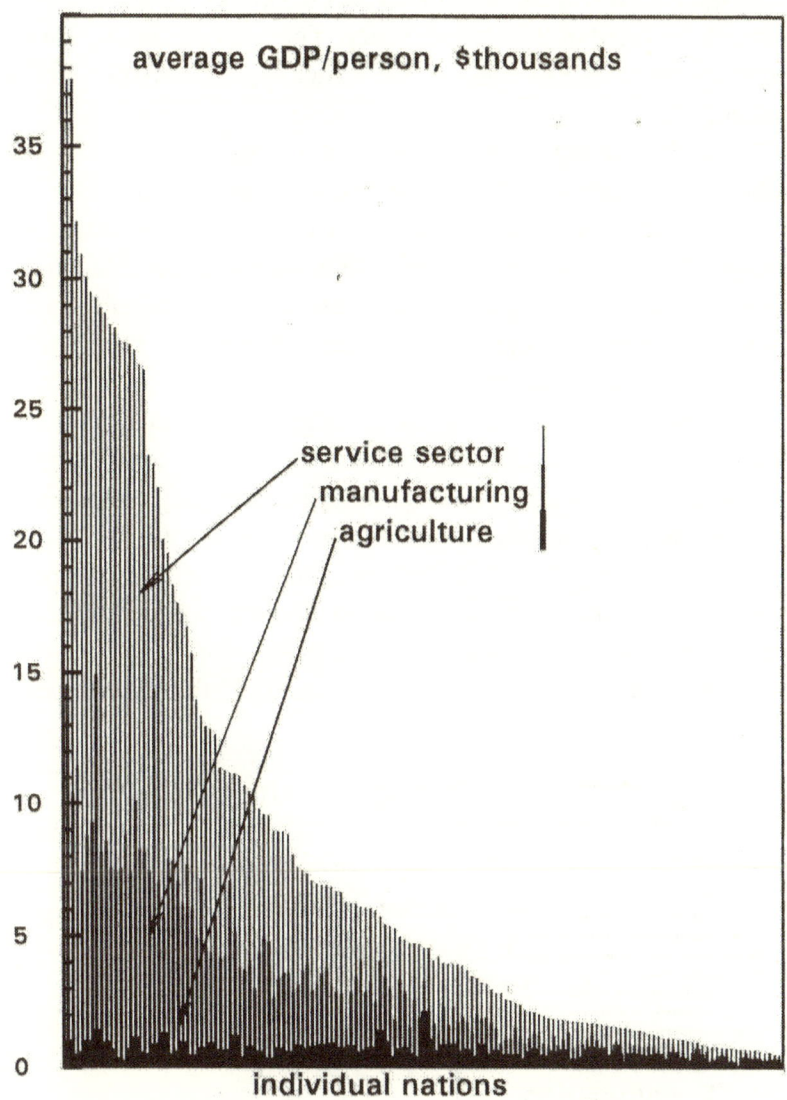

Fig. 1.2 Contributions to the average GDP/person

Data source: Central Intelligence Agency, Factbook 2004

Fig. 1.3 shows the occupational diversity in relation to the sectors of the GDP of the United States. The light and heavy bars are the change from 1939 (light) to 1997 (heavy). The arrows indicate the major trends.

The decreasing fraction for food is consistent with higher productivity in farming. It includes the increased fraction for food processing. The decreasing fraction for clothing is consistent with higher productivity in manufacturing. This includes the effect of free trade with nations having lower cost labor. The increasing fraction for health care reflects the higher demand that results from success in preventing premature death.

Technologies such as nanotechnolgy, proteomics, photonics, superconductivity, and informatics were unknown a generation ago. They already have secure places among the emerging technologies.

The major common denominator is electric power. It is too pervasive to specify why this is so in detail. It magnifies the power a human can exert. It amplifies the versatility, speed, reach, and/ or precision many tasks demand. It creates a work environment in which people can be effective. It provides tools that enable people to use time more effectively.

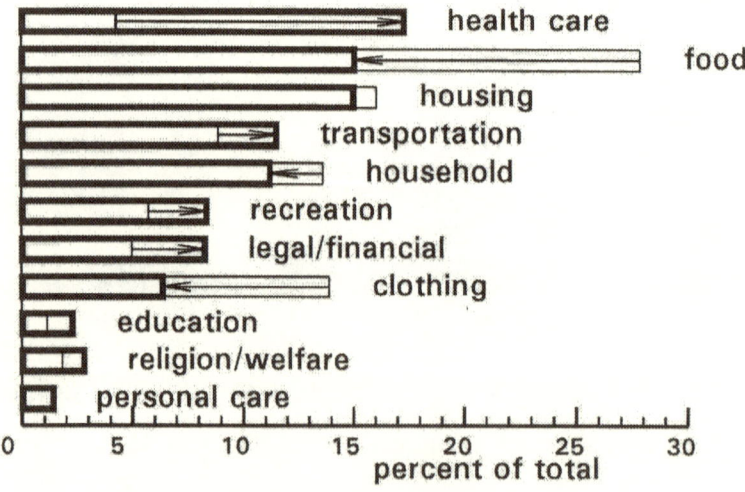

Fig. 1.3 U.S. Occupations 1937-1997

Data source: National Income and Product Accounts 1999

National Responsibility

As occupations grow more diverse trade becomes more necessary. One of the first duties of government is to establish a national currency. Money is a token representation of value used for trade. The government owns it and loans it to banks for use in trade. The value of the goods and services it can buy determines its value.

Monetary policy maintains the value of money. It is a major government responsibility[14]. The government fails if the value of its money fails. The value of money is stable when the amount of money in circulation balances the needs of current economic activity. A *tight money policy* slows economic activity. A *loose money policy* increases it artificially causing rising prices. Governments usually allow small, but finite inflation. This promotes strong economic activity, causes only minor uncertainty, and gradually devalues past debts and unproductive assets.

Capitalism provides a way to spread the cost of the expensive tools of a large enterprise over a long time period. The cost effectiveness of very large steam engines caused capitalism to flower three centuries ago[15, 16].

Private investors form a corporation. They buy shares of the corporation to finance the large enterprise. They receive a return on the investment from the income it generates and from increased value of the shares if the venture is successful. They also risk losing the investment if the venture is not successful. Electric power originated as a private enterprise.

The obvious disadvantage is that successful ventures displace workers who must find other employment. Tension between the conspicuous wealth of the most successful capitalists and lowest income workers is not new. It is exacerbated by the increasing pace of progress. The largest corporations can no longer guarantee employees lifetime security. When new technology displaces employees they may need not only new employment but new careers. Government ownership of productive assets does not appear to be an answer. Social security is a continuing unresolved dilemma.

Fig. 1.4 shows the U.S. fiscal policy and features of its dynamics. The numbers show the creation and distribution of the national GDP for 2005. Public investment, private investment, and social investment are three major feedback loops that sustain the ability to earn income. The main difference among nations is in how their policies balance the three loops.

The tax revenues have been about 20% of GDP for the past half century. Governments control tax rates. Tax revenues depend on GDP as well as tax rates. The effect of tax rates on GDP is not transparent to the layman. Politically tax rates must appear progressive and equitable. Rates that produce the greatest GDP must balance the needs of government.

Social investment helps cope with the economic dislocation inherent under capitalist free trade. It also provides a safety net for people with misfortunes they cannot handle themselves. Private insurance, entitlements, and welfare serve overlapping purposes.

Public investments are government services that establish a productive environment. They include public education, the monetary system, the legal system, the regulatory systems, international affairs, military protection, and the infrastructure for transportation.

Private investment creates businesses. The labor force earns roughly two thirds of the income. The other third is return on investment. It pays for creating and financing the enterprises that range from single owner proprietorships to large corporations. Much of the return on investment becomes new investment.

Private wealth is mostly residential housing. The desire to own property is historic. Ownership of land and its association with food production was once the major source of political power. The sources of political power grow more diverse as the ways people earn income grow more diverse.

Foreign exchange is the unbalance between international trade and international investment. U.S. currently has a large foreign trade unbalance due largely to low labor prices in the emerging economies. The *capital account* surplus is mostly investment in U.S. enterprises by people in foreign countries who own U.S. money. It must balance the *current account* trade deficit.

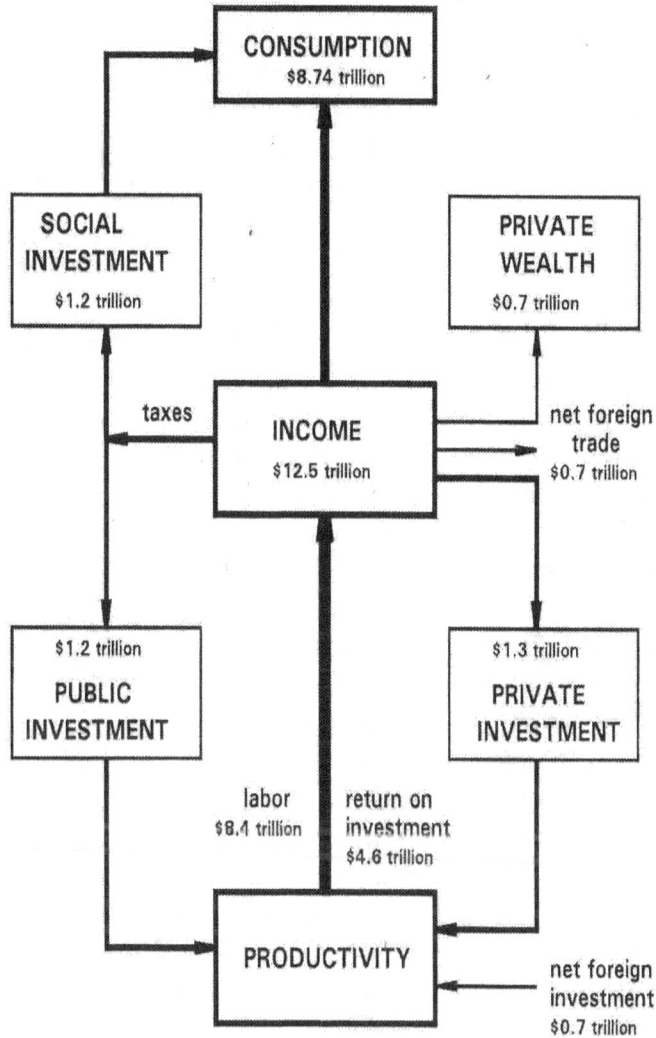

Fig. 1.4 Schematic U.S. 2005 fiscal policy dynamics

Data source: National Income and Product Accounts
U.S. Department of Comerce Bureau of Economic Analysis

World Responsibility

The Bretton Woods Conference during World War II met to develop a plan to avoid the mistakes in recovering from World War I[17]. Representatives of 45 nations proposed *global free markets* as an alternative to imperialism. The premise is that individual productivity is the basis of all wealth. Individual ingenuity and organization of assets are inherently unlimited. Natural resources support individual productivity in the nation that owns them.

Great Britain abandoned the greatest empire in history. It became the United Kingdom. The United States abandoned its traditional isolation from World politics. The conference created antecedents of the *International Monetary Fund,* the *World Bank,* and the *World Trade Organization.*

The immediate concern was to protect the currency of nations emerging from the chaos of war. The international monetary fund set financial policies nations desiring protection must satisfy. These policies also attract private investment. This has been the basis for world economic development ever since.

Free trade and investment is between individuals of any nation. Nations benefit from the economic activity and income. Creative individuals, corporations, and governments profit from trade. People in an importing nation buy goods and services in its own currency. The exporting nation must spend the same currency. It generally does this by investing in enterprises of the importing nation[18]. Currency exchange markets set currency at a value that balances the import expenditures and investment receipts.

International trade creates global markets. It does not have to be a large fraction of domestic trade to do this. Prices for labor are local and lag other development. Underdeveloped nations with a potentially productive work force have low labor prices. This trade advantage is a temporary window for development.

Economic parity exists when labor, goods, and services all have global market prices. Parity is a condition the world international economic system can create. Further development depends on whether individual nations develop competitive policies and people.

After World War II the U.S. with 5% of the world population was the only wealthy developed nation with an economy that was intact. About a third of the world's 6 billion people now live in developed nations.

Japan has half the U.S. population. It succeeded in showing that global free trade can overcome overpopulation and inadequate natural resources. The U.S. work force and economy adapted to the competition with considerable dislocation.

Underdeveloped nations have three times the population of developed nations. China's population alone is double that of developed nations. Capital is now flowing from many sources. The impact of the increased productivity of its low cost labor is obvious. The population of India is almost as large. Absorbing the productivity growth of this much low cost labor is a serious challenge for the labor force of all developed nations together.

The State of World Development

Fig. 1.5 shows the state of world electric power[19]. Public lighting is a small part of electric power. It is a visual indictor of the state of the world. The success of the Breton Woods strategy is apparent. Night activity highlights centers of trade and transportation. It identifies the location of economic activity. It does not directly show wealth. The average illumination of India, one of the poorest nations, is only moderately less than the average illumination in Europe, the world's richest nations.

Electric power began in European industrial centers familiar with large steam engines. The lights trace the spread to other centers of trade and transportation. The large dark areas are land that is not arable and underdeveloped nations. Much inland area with scattered lighting is arable land.

The Ural Mountains and the Caspian Sea inhibited the spread of electric power eastward. Harsh climates inhibit the spread to Siberia and Manchuria. The Gobi Desert is a barrier southward to Kazakhstan, Mongolia, Tibet, and Western China. Industrialization of the Far East began in Japan. It spread south and west.

13

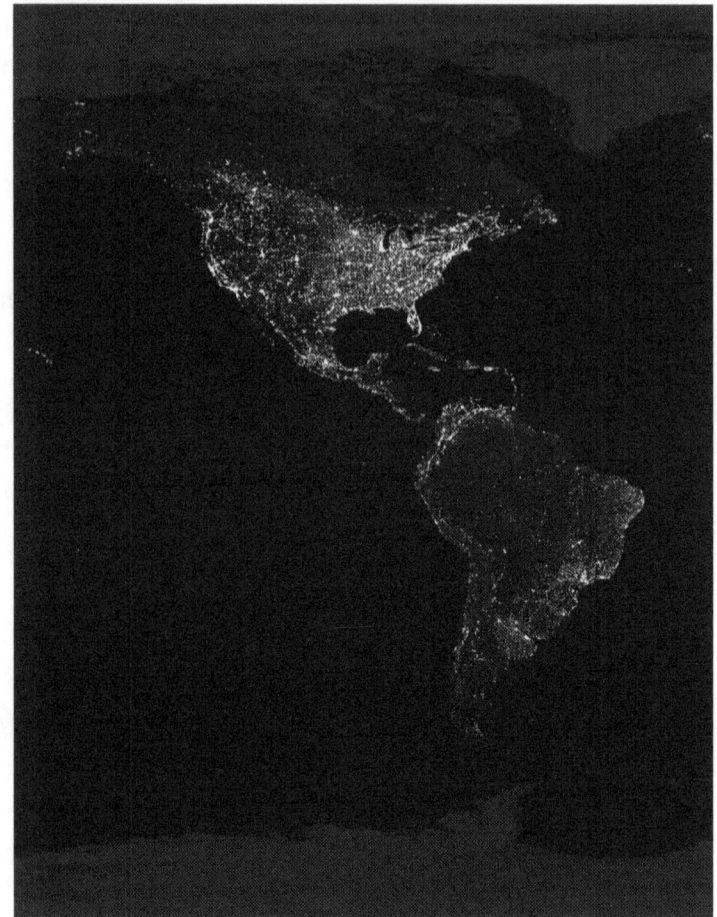

Fig. 1.5a Western hemisphere of the earth at night

Fig. 1.5 is a composite satellite photograph showing the light emitted by population centers in all nations with an electric power infrastructure. All major cities are easily identifiable. The size of each point represents some combination of urbanized population and productive intensity. The density of points is greatest in coastal regions where international trade activity and international standards of developed nations prevail.

Fig. 1.5b Eastern hemisphere of the earth at night

Africa and South America have significant populations that do not depend on commerce. Australia and Siberia are areas with low population density. The illumination follows historic trade routes through some of the harsh climates, such as the ancient silk route to China through the southern nations of the Russian Federation, the Trans-Siberia railroad through the harsh climates of Siberia, and the Alaskan highway through Canada.

Notes and references for chapter 1

1. Continuous changes in mitochondrial DNA give people their individuality. Over time advantages in the ability to survive re-enforce some of the DNA changes giving a survival advantage to individuals with specific traits. Over many generations these can cause a divide in the DNA record. Specifying which break marks a change in species is somewhat arbitrary. The migration out of northern Africa 100,000 years ago is widely accepted as the divide between modern and pre-existing humans. Richard E. Leakey and Roger Lewin, *Origins reconsidered: In search of what makes us human*, Doubleday, New York 1992

2. DNA evidence for the date of origin of modern humans D.B.Goldman, et. al. *Genetic absolute dating based on micro-satellites and the origin of modern humans*, Proceedings of the U.S. Academy of Sciences, **92**, 6723, 1995

3. The major Polynesian islands, Australia, and New Zealand were originally populated about 40,000 years ago. About 6,000 years ago people on the coasts of Polynesia were supplanted by Chinese from Taiwan, leaving an indigenous population in the central jungles. They were probably related to the Maoris of Australia and New Zealand. Geoffrey Irwin, *The Prehistoric Exploration and Colonization of the Pacific*, Cambridge University Press, 1992

4. One origin of the Western hemisphere population is dated at about 14,000 years ago by the Clovis hunting tools used to hunt wooly mammoths, possibly to their extinction. Normal migration accounts for extending the population to Argentina within a millennium. Gary Haynes, *The Clovis era in a mammoth haunted continent,* Cambridge University Press, New York, 2002

5. The archeological evidence of cultural advance starting about 50,000 years ago is based on artifacts such as cave art. Richard Klein and Blake Edgar, *The Dawn of Human Culture*, John Wiley, New York, 2002

6. Dean Fraser, *The People Problem*, Indiana University Press, 1971

7. Joel E. Cohen, *How Many People Can the Earth Support*, W.W. Norton & Company, New York, 1995

8. Reasons for the pattern of origin and spread of global agriculture have been derived from the anthropology of different people and the agriculture they have developed. They are presented and documented with an extensive bibliography Jared Diamond, *Guns, Germs, and Steel,* W.W. Norton, Inc, 1999

9. Farm lands reduce bio-diversity with single species crops. Farm lands that are marginal produce more soil erosion than more productive land. A case is made that farming land more marginal than that now being cultivated would cost more in erosion and lost bio-diversity than would be gained. Dennis A. Avery, *Hudson Institute Reports*

10. Note that production per farmer is not the same as production/acre. Chinese farmers produce several times more per acre than the U.S. production/acre to feed the population. They do this by growing several crops per year.

11. Data for the different geographic regions is a weighted average taken from the 150 largest nations. U.S. Central Intelligence Agency *Factbook 2004*

12. Defeating the combined forces of enemies on two fronts during World War II depended on the superior productivity of the United States. The GDP concept was developed at the *U.S. Bureau of Economic Analysis* to determine the limits of that productivity. Professor Simon Kuznets, received the Nobel Prize in economics for this work. It is updated monthly in the *National Income and Product Accounts,* compiled by Department of Commerce in *Survey of Current Business.*

13. The purchase price parity method of calculating the GDP gives the value of goods and services would have in dollars in U.S. markets. In developed nations this is essentially the same as their value in local markets converted to dollars by the currency exchange rate. In underdeveloped nations it is often higher, particularly in survival goods such as food. *Review of ppp-adjusted GDP,* International Monetary Fund Working Paper WP/95/18.

14. The first goal of *monetary policy* is to maintain confidence in the value of money. Federal Reserve Banks manage U.S. monetary policy as a private corporation under government oversight. The Federal Reserve Banks put money in circulation by loaning it to commercial banks at a low *federal funds rate.* They control the amount of money in circulation by the interest rate and by sales of Federal Reserve Securities. They also control business activity by making their intention transparent to the business community.

15. Oxford University Professor Adam Smith (1763-1776) gave the definitive description of capitalism in lecture notes which have been republished with commentary by Edwin Cannan. Adam Smith, *The Wealth of Nations* Random House Modern Library Series

16. Federal Reserve Commission Chairman Alan Greenspan in testimony before the Joint Budget Committee of the U.S. Congress, June 2005, commented that Adam Smith's notes remain a modern economics treatise.

17. The *Bretton Woods Conference* met for 4 weeks in July 1944 just after the Normandy invasion of World War II. The agenda was an Anglo-American initiative to address problems that would arise at the end of hostilities, specifically, the devastation of productive capacities and the impact of war debt on the value of national currencies. The conference established the *International Monetary Fund* (IMF) to provide liquidity in a currency exchange crisis. It established the *International Bank for Reconstruction and Development (World Bank)* as a source of loans and grants for development. It established the *General Agreement on Tariffs and Trade* (GATT) which evolved to become the *World Trade Organization* to reduce barriers to free trade. Harold James, *International Monetary Cooperation Since Bretton Woods*, Oxford Press, 1996

18. Nations have a *foreign account* that consists of two parts. The *current account* is the balance of its own money it sends to other nations by buying imports and the foreign money it imports by selling exports. The *capital account* is the balance between its own money it exports by investments in foreign assets and foreign money it imports by foreign investments in its assets.

Currency exchange markets continuously buy and sell currency to change prices to reflect the market estimate of real values. Ultimately this balances exchanges of currency of each nation to zero. The value of money then *floats* at the international exchange market price.

19. The imagery for Fig. 1.15a and Fig. 15b as well as the book cover was produced by the NASA Visible Earth Team and displayed on the internet as http://visibleearth.nasa.gov\.

Chapter 2
World Energy Demand

An electric power grid combines the power of steam engines with the versatility of electricity. It gives connections to power almost everywhere any time at low cost. The history of the grid began a century ago. The energy source was primarily coal. The coal supply is very large but not infinite. The growth in demand is large enough to expect changes well beyond an extrapolation of recent experience. This chapter examines the past growth of demand and factors that will make the future different.

Origin of Electric Power Demand

Industrialism and capitalist enterprise evolved throughout the 18th century. Manufacturing took the form of smokestack industries. They relied on the efficiency of scale of large central steam engines to power all the machinery in a factory.

Magnetic induction is the force field an electric current produces. Faraday in England and Joseph Henry in America discovered it in about 1830. Science soon learned how to apply it. The watt, joule, ampere, volt, ohm, farad, hertz, and tesla are units of electromagnetic measurement named for major contributors.

Electromagnetic force can couple one mechanical motion to another. A steam engine can drive an electromagnetic generator. This produces an electric current. The current produces the electromagnetic force that drives motors. The system converts mechanical power to electrical power with very little loss[1]. Electric power transmission is simpler than mechanical transmission belts, pulleys, and drive shafts. It makes the power of the steam engine available at great distances with another minor loss.

The tungsten light, invented by Edison late in the 19th century, was a convenient replacement for more cumbersome carbon arc lanterns. It was the incentive for universal distribution[2]. By the end of the 19th century the Edison Electric Power Company delivered power to about 100 New York mansions.

A safe residential voltage is 100-200 volts. An ordinary wire carries currents up to about 50 amps. This is enough current for a residence at a safe voltage but not for number of houses or heavier power users. This requires much higher voltage to stay within exceeding the current capacity of an ordinary cable.

Transformers use alternating current to trade high voltage for current with very little loss in power. They deliver whatever power users require at a high voltage, and then transform it to a safe voltage[3]. The ac motors and generators eliminate high maintenance contacts between the rotor and stator coils of dc machinery.

Alternating current power was an obsession of Nikola Tesla. He patented the main ideas for two phase and poly-phase alternating current generators and motors while employed by Edison's telephone and telegraph office in Paris[4]. On emigrating to the U.S. in 1884 he worked at miscellaneous electrical engineering projects as an Edison associate. Edison did not have Tesla's scientific education and background. He did not understand or declined to admit the advantages of alternating current. Tesla left his association with Edison to pursue his ideas. He eventually got the financial support and friendship of George Westinghouse, the inventor of a failsafe railroad braking system. Meanwhile the Edison Company became overextended and was forced into a merger that ultimately became the General Electric Company.

A controversy over whether the power for the grid would be from alternating or direct current generators followed. The Edison advocates electrocuted animals with ac current to demonstrate its danger. Tesla staged spectacular electrical demonstrations to show its safety. Tesla and Westinghouse ultimately succeeded in winning the contract to generate and distribute hydroelectric power from Niagara Falls. This assured that future power distribution would be based on alternating current technology.

Electric lighting was the decisive impact that made wide distribution of electric power to individuals a commonplace necessity. Population growth stimulated additional distribution growth. Diversified employment and applications that increased the growth of individual productivity quickly followed.

Tall smoke stacks characterized manufacturing of the mid 19th to the mid 20th century. Coal fired furnaces produced steam. Steam engines generated power. A collection of rotating shafts and belt drives transmitted power throughout the factories. They drove conveyor belts, fabrication machines, power tools, and manufacturing equipment of all kinds. The central power source did the heavy lifting. The operations were labor intensive and not highly automated.

Central steam engines centralized labor. Large factories required large numbers of employees. Since the first electric power plants used the same energy sources and engines as manufacturing, the economics of electrical power paralleled the economics of steam power. The unit cost of power decreases with the scale of the power plant up to a gigawatt.

From the start commercial electric power relied heavily on the huge engines that drove the industrial revolution. Expansion to a nationwide power distribution grid made electricity universally available to all parts of the country at low cost. Electricity gave everyone from individuals to major industries access to power with little or no capital investment. This meant broader access to ways to be productive in small businesses with modest financing.

Decentralized population and employment is a growing consequence of a combination of electrical power, transportation, and communication. Communication is making decentralization global. Greater specialization requires more services. Electricity gives everyone from individuals to major industries access to power in any amount with little or no capital investment. This means broader access to ways to be productive in small businesses with modest financing.

Transportation and communication decentralize employment and population further. The entire world population is affected in some way. The developed part of the population of each nation is global in its outlook and dependence. The underdeveloped part of the population of each nation is tribal in its outlook. The thrust of current change is still global. The dichotomy that remains among both individuals and population classes limits the rate of change.

Electric Power and GDP

Fig. 2.1 compares electric power with GDP in the world geographic regions[5]. The vertical bars represent population. The base of the diagram plots the GDP per person as a function of electric power consumption per person. The diagonal line across the base of the diagram has a slope of $25,000 of individual GDP per kW of electric power per person.

The correlation between individual productivity and electric power consumption is clear. The power consumption in dollars gives a clearer picture of its nature. At an average price of 10 cents per kWh the annual cost of 1 kW of electric power consumption is $876 per person. The average annual GDP/person is $25,000. The national income is 28 times the cost of the power a nation consumes.

The correlation shows that electric power has an essential connection with average individual productivity. The cost of power is less than 4% of the income due to productivity. Some uses of power contribute directly to productivity and GDP. Some contribute to comfort and standard of living. Some uses are frivolous.

Nations deviate from the average for individual reasons. Norway, Sweden, Finland, and Canada have high consumption per person partly due to low cost hydroelectric resources. Kuwait and the United Arab Emirates have low cost petroleum. These factors plus other differences in culture and development do not obscure the firm relation of electric power to economic productivity.

The economic power of nations and regions is the total GDP. This is the GDP/person times the population. Europe, North America, and East Asia are by far the dominant world economic powers. Their GDP is $11, $12, and $13 trillion, respectively. The large economic power of East Asia and India is due more to their large populations than a high individual productivity. Economic development in regions which have large populations has a disproportionate impact on world energy resources and economic power. Enough development is already apparent to assure that the Indian Sub-continent and East Asia will be much more dominant world economic powers of the future.

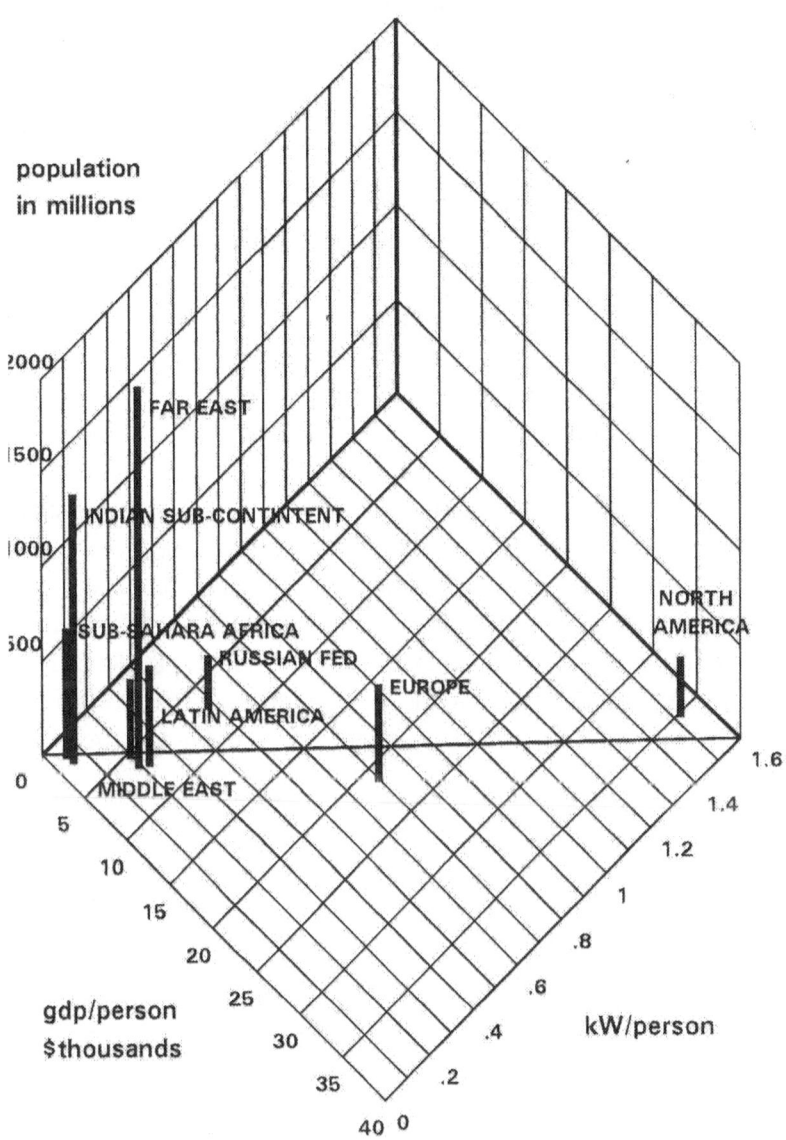

Fig. 2.1 Individual productivity and electric power.

Source: Central Intelligence Agency, Factbook 2004

Consumption of Energy for Electric Power

Fig. 2.2 shows how the primary consumers use energy during the past half century in the United States. A primary consumer of coal is the user who purchases the coal from the mine. Electric power producers are primary consumers of coal. Electric power consumers are not primary users of the same coal.

Note that the GW units of this figure represent fuel energy that is heat. The heat used to produce electric power is 3 times the electric power it can produce.

The narrower lines show the relative growth of population. Over the 50 year period shown the population grew by 181%. Over the same period the GDP/person grew by 2800%. This is far off the scale of this graph. Industrial, commercial, and residential heat consumption increased in proportion to the population.

Electric power has grown at a rate 2.5 times faster than transportation. This contradicts the impression people may have from the more visible traffic growth on freeways and from fuel-consuming airline travel.

Table 2.1 compares the growth in energy consumption in each sector as a multiple of the population growth and as a multiple of the GDP/ person. This shows little change in the individual demand for residential, commercial, and industrial heat except possibly more efficient use.

Table 2.1 Growth of U.S. energy, population, and GDP/person

	Energy growth	energy growth / pop growth	energy growth / GDP growth
Electric power	8.18	4.52	0.28
Transportation	3.16	1.75	0.11
Industrial heat	1.64	0.91	0.06
Commercial heat	1.51	0.84	0.05
Residential heat	1.49	0.82	0.05
Total	2.86	1.58	0.10

Data source: U.S. Central Intelligence Agency, Factbook 2004

The growth of electric power consumption dominates that of other energy uses. This is both as a fraction of GDP growth and as a multiple of population growth.

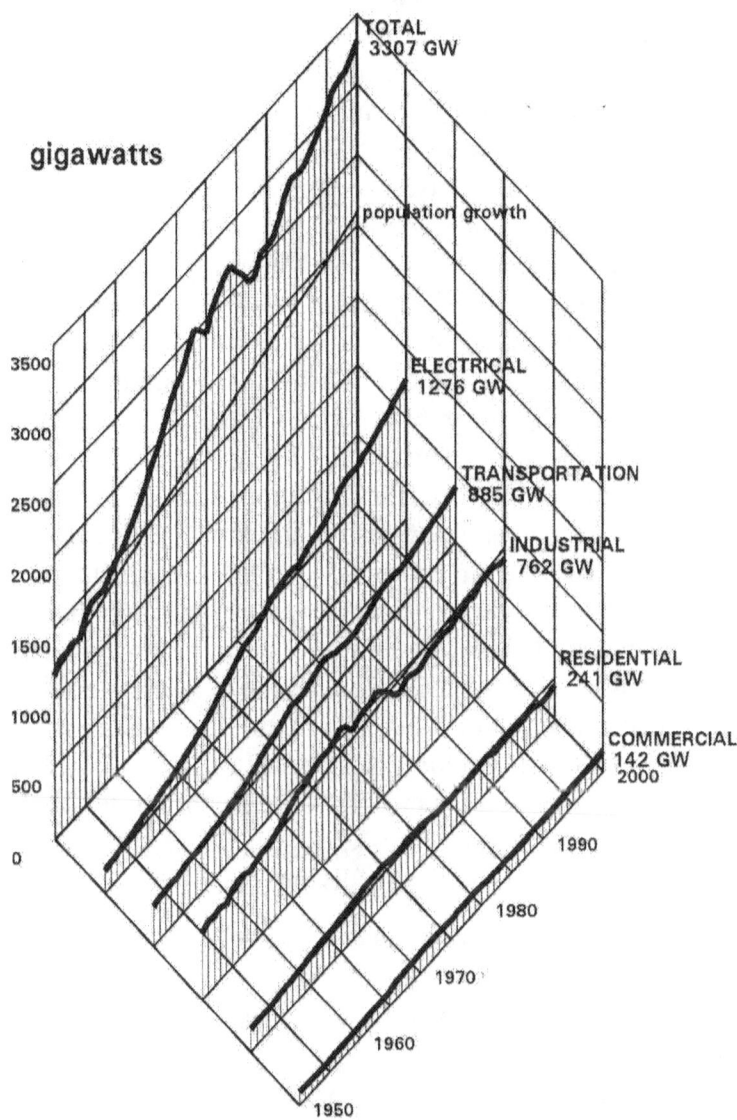

gigawatts

Fig. 2.2 U.S. primary energy consumers.

Data Source: Annual Energy Review 2000, U.S. Energy
Information Agency, U.S. Department of Energy.

Fig. 2.3 shows World primary energy demand among different users. It uses the same 5 decade scale as the U.S. data for comparison. However, reliable world data have been available for only the two decades since the end of the Cold War. As before, primary energy for domestic use, industry, and commerce is mostly as heat. It excludes electrical energy, which is a secondary demand. Worldwide the major energy use is for domestic heating and cooking. That demand increased, unlike the U.S.

Table 2.2 compares the U.S. and World demand for energy expressed as watts/person, the percent for different uses, and the growth over the past two decades. Except for domestic use, the pattern of demand and the pattern of growth are similar to the U.S. In spite of the huge discrepancy in the magnitude of the demand per person, electric power accounts for the greatest growth in demand and a similar rate of growth. Transportation accounts for the next greatest growth in demand and also a similar rate of growth. Per capita industrial use of primary energy is roughly constant.

The world is still largely underdeveloped in comparison with the U.S. The underdeveloped nations have greater growth in demand for electric power and transportation than the U.S. Their energy use for power and transportation purpose is still a smaller percent of the total. The U.S. has the highest standard of living by the GDP and individual productivity standards. Whether or not the U.S. is a model to follow, it is important to know what energy consumption has to do with GDP and standard of living.

Table 2.2 U.S. and World per capita energy use

	United States			World		
	W/person	pct	growth	W/person	pct	growth
electric power	4640	39	1.57	535	25	1.67
transportation	3218	27	1.35	348	16	1.48
industrial	2721	23	1.01	396	19	0.99
domestic	876	7	0.96	544	25	1.27
total	12025	1.26		2116	1.38	

Data Source: World Bank, World Development Report 2000

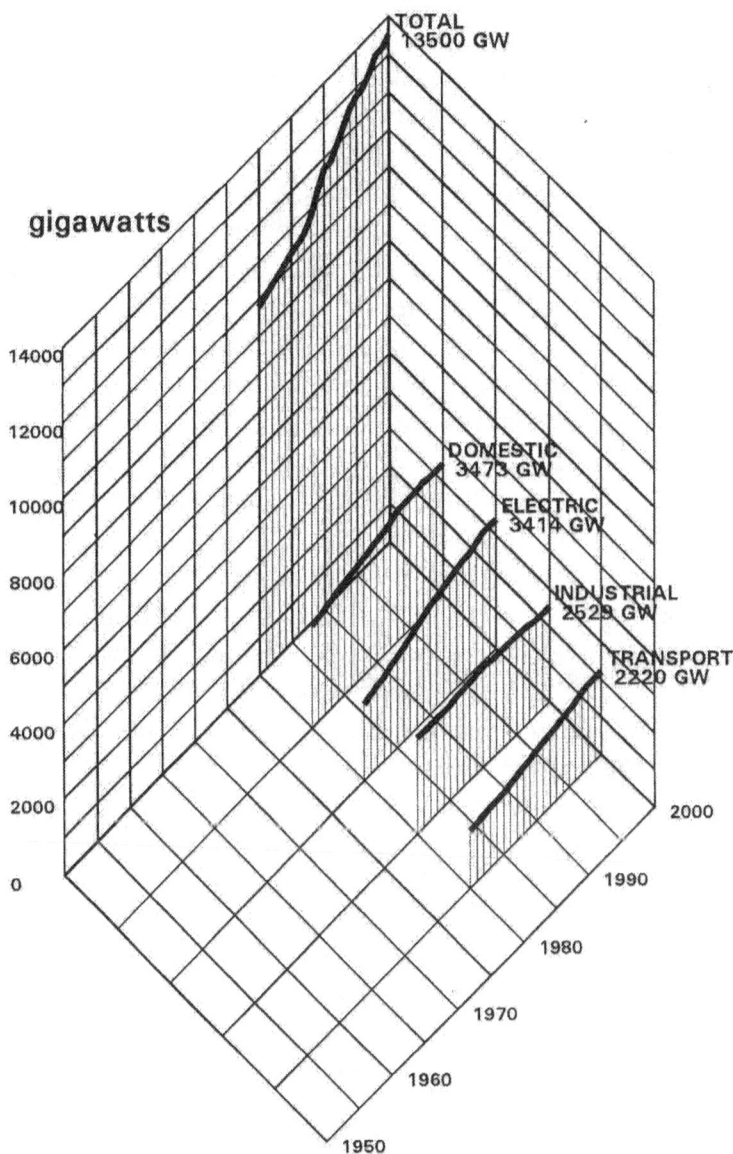

Fig. 2.3 World primary energy consumers

Data Source: British Petroleum Company.

Fig. 2.4 shows electrical power demand by consumers in the United States over the past half century. This is the electric power component of the heat demand in the previous figures. Each class of heat consumer consumes electric power in addition to heat. The factor of three heat-to-electricity conversion is included in the electric power that industry consumes.

Electric power users are divided roughly equally among residential, commercial, and industrial users. Local governments use most of the small remainder for functions such as street lighting, water supply, and sewage disposal. Differences in the rate of growth of residential, industrial, and commercial use are insignificant.

The correlation between electric power and the GDP is reasonably clear from the comparison of widely varied geographic areas. No correlation with the growth of electric power use by the sectors that contribute most directly to GDP is apparent. The growth of GDP far exceeds the growth of electric power. Whatever the cause, the effect of electric power on GDP is increasing.

Large motors give industry the flexibility to provide mechanical power where it is needed to improve productivity through automation. Since few factories can compete with the efficiencies of scale of a large power plant they purchase power as a direct cost of manufacturing productivity.

Air conditioning, temperature control, and air circulation accounts for the large peak demand in summer and a smaller peak in winter. It is an indirect cost of productivity of people who have it in their homes as well as the workplace.

Many technologies, such as communication, information technology, and remote sensing automation, depend on electric power but have relatively small demand for power. These functions are consistent with growth of electric power and much larger growth of GDP.

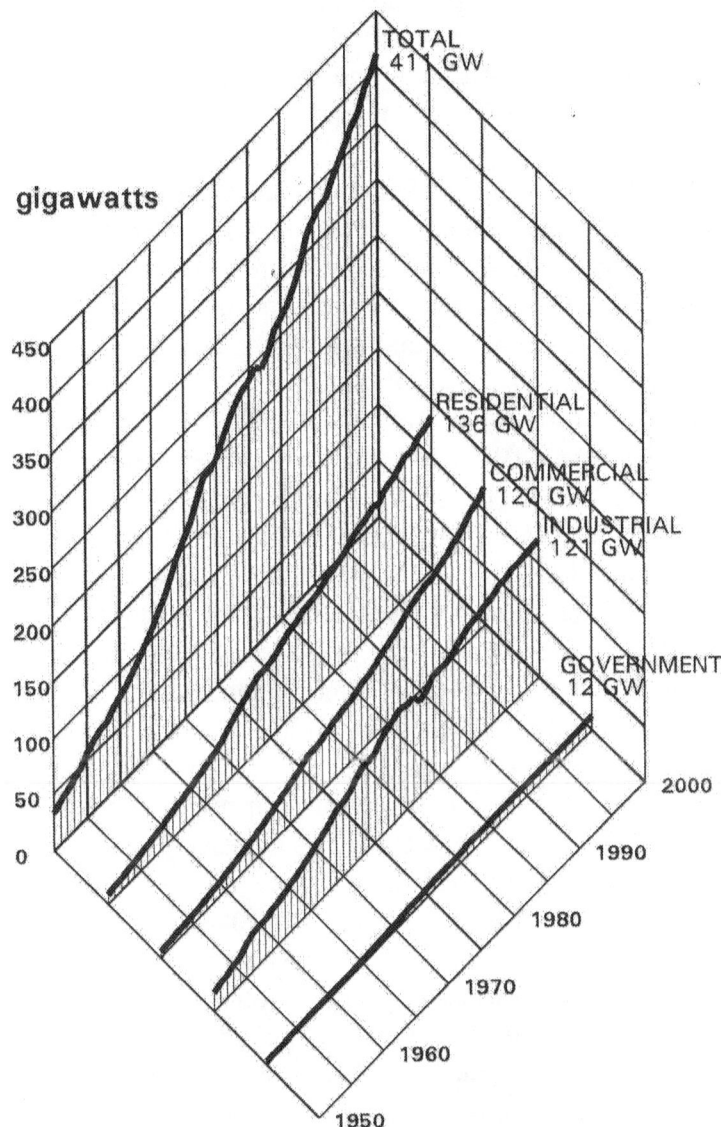

gigawatts

Fig. 2.4 U.S. electric power consumers

Data Source: Annual Energy Review 2000, U.S. Energy Information Agency.

Energy Sources for U.S. Electric Power

Fig. 2.5 shows the past history of the use of fossil fuels that supply most of the 400 GW of electric power that is currently consumed in the U.S. Although the cost and availability of fuel is the subject of Chapter 3, it also depends on the history of electric power demand.

Coal provides more than half the electrical power. Like nuclear power it is available in abundance. It provides power with high reliability. Economies of scale keep the cost of power low. They limit the flexibility to change the power output level. Large deviations from average demand must be met by more flexible, but more expensive gas.

Nuclear power has grown steadily since about 1970. It accounts for about a 19% of the total and is the primary alternative to coal for base power. The scale of nuclear power plants is large for technical reasons. Its output level is inflexible. The economies of scale are similar to coal.

Natural gas grew to about 25% of the total in 1970 with the growth of pipelines and after 20 years of static demand it has again grown to about 15%. Its reliable versatility in supplying peak demand power, and the avoided cost of major capital projects, make gas useful in spite of its higher cost.

Petroleum reached a maximum contribution of about 20% in the 1970s and has declined ever since as increasing cost outweighed its convenience. It is no longer used for new power installations due to its higher priority use for transportation. That competition raises the price of oil.

Hydroelectric power depends on other demands for water and on annual rainfall. The best hydroelectric resources are now developed. Their contribution peaked at about 25% over 20 years ago and has decreased to less than 10% of the total.

Renewable sources have grown to about 2% of the total. Hydroelectric power is a model for the magnitude renewable energy sources must reach to become significant. The limited supply and uncertain reliability places restrictions on how they can be used.

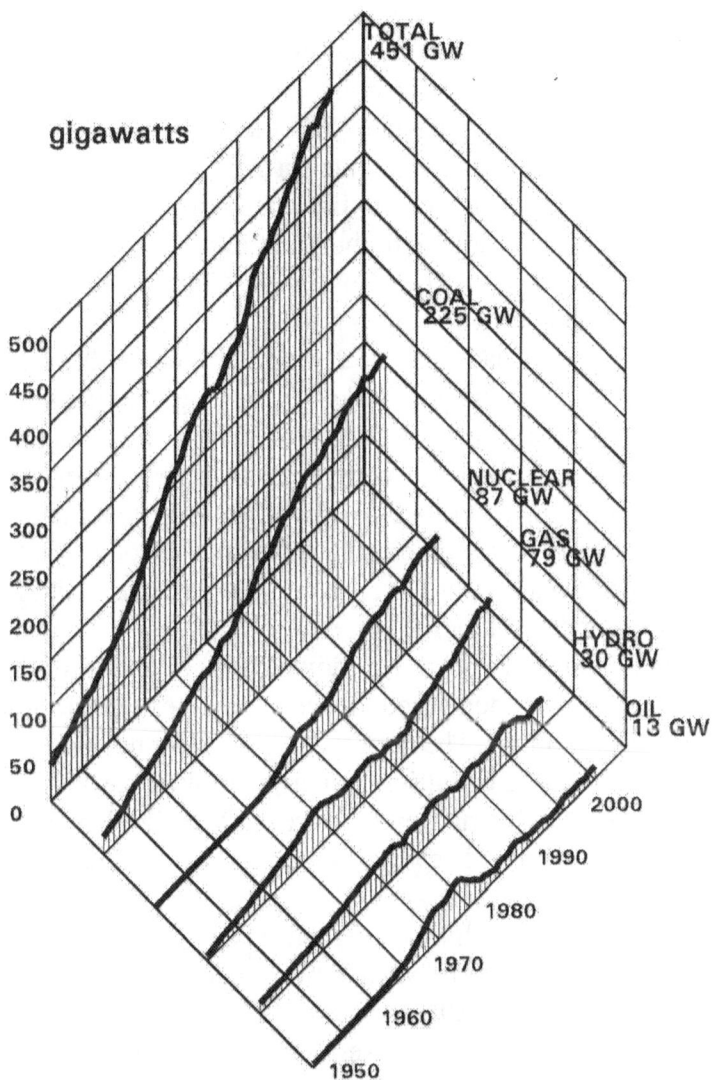

gigawatts

TOTAL
451 GW

COAL
225 GW

NUCLEAR
87 GW

GAS
79 GW

HYDRO
30 GW

OIL
13 GW

Fig. 2.5 Sources of U.S. electric power

Data Source: Annual Energy Review 2000, U.S. Energy
Information Agency

World Fossil Fuel Consumption

Fig. 2.6 shows how the world energy demand is divided among users of carbon based fuel.

Petroleum accounts for the largest use of energy. A major fraction is used for home heating and cooking. Worldwide transportation accounts for less than half the total in contrast with the U.S. where transportation accounts for nearly 3/4. Oil is by far the largest commodity exchanged in international trade.

Coal deposits are large and abundant in a few of the largest nations. The largest are in the U.S., Russia, and China. In comparison with oil, international trade in coal is small.

Natural gas has had fast, steady growth due to the growth of pipeline and, more recently, liquid natural gas technologies. In contrast with oil, a large percentage of the gas in a deposit is recoverable. No secondary recovery methods can recover more. Furthermore, some gas deposits provide the pressure that makes an underlying oil deposit recoverable. Technology to produce natural gas from the deep ocean clathrate formations could make natural gas the most abundant carbon fuel.

Nuclear and hydroelectric power makes the difference between the total and carbon based fuel consumed for electric power generation in most nations. The increased demand for nuclear power is particularly strong in countries that have limited carbon based fuel resources. Hydroelectric power sources of low cost power are mostly already developed. They meet an important but small part of the total demand.

Japan is proof that human ingenuity and productivity can produce prosperity in nations with limited natural resources and large population densities. Although Japan is a large nation, it is not large enough for its demand to be a serious test of the limits of world energy supplies. The real test will come as China, India, and other nations with high populations press the advantage of their low labor cost as they become more prosperous.

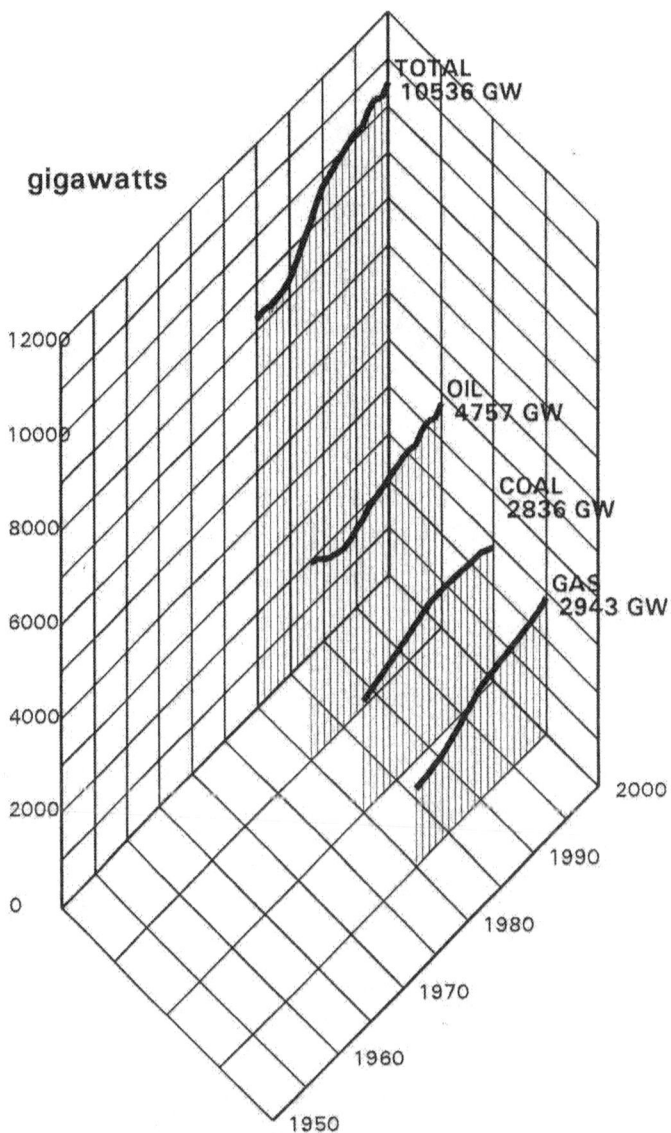

Fig. 2.6 World fossil fuel consumption

Data Source: British Petroleum Company.

Estimating Future Demand

Table 2.3 compares the present world demand for electric power with the possible consequences of economic development. The last column shows the demand if average GDP/person is equal to the present average in North America. This assumes that electric power demand will increase in proportion to GDP. The bottom line world increase is a factor of 5.5 by underdeveloped nations alone. This is not a prediction or even a good basis for a prediction. It is only a look at one factor. An alternative is to express the world demand in relation to an arbitrary standard demand.

Table 2.3 World energy consumption

	pop millions	gdp / person	W/person electricity	present total GW	ultimate total GW
North America	347	$34548	1487	516	516
Europe	520	21336	694	361	773
Russian Federation	285	7122	477	136	424
Latin America	542	7042	216	117	805
East Asia	2033	6538	187	380	2990
Mid-East	429	5452	189	81	637
India	1428	2368	50	72	2141
Sub-Sahara Africa	693	1622	46	32	1034
World	6302	7775	269	1695	9370

Data Source: World Bank, World Development Report 2000

A standard 1 kW/person is easy to remember. The demand by any sector of the population is the standard demand times the population. Each factor that affects the demand is a multiplier of the standard demand. North American consumption is 1.48 times the standard. World consumption averages 0.26 times the standard. A practical future is the 3-decade life of a new power plant. The arbitrary working assumption is that a 0.5 kWh/person target world individual consumption and a 5000 GW_e total demand are realistic. Future growth or decline of electric power consumption depends on many factors such as population, economic development, technology change, productivity change, fuel cost, and conservation.

The factors that increase or decrease the demand are not independent. Future demand depends as much on the interactions as the factors themselves. There are obviously dozens of interactions. Three examples illustrate the kind of thinking that follows from a particular choice.

The relation between population growth and GDP is clear. The nations with the highest population growth have the lowest GDP/person. The success of medical science in extending the average life expectancy requires a compensating control of birth rates. Competitive individual productivity depends increasingly on knowledge and education. The difficulty and cost of assuring that offspring are competitive appears to deter excessive birth rates. In spite of its obvious association with ignorance and poverty high population density does not. In China and India the response to high population density was more intensive agriculture that maximizes the yield-per-acre by growing several crops per year.

The benefits of knowledge require increasingly more versatile ways to take advantage of it. The important attribute of electric power is versatility. Recent history indicates that the demand for electric power will continue to increase as long as inexpensive power is available. Price incentives can and should encourage conservation through efficient use of power. However the demand for electric power may be a factor driving population toward a stable steady state.

Many large nations are progressing toward economic parity based on the high productivity of relatively low cost labor. For economic parity the price of a labor market must reach the global market price for labor. For full economic equality the knowledge of the labor force must be competitive. Japan is an overpopulated nation with limited resources. Like other developed nations it progressed beyond economic parity to economic equality, if not superioity. China and India have huge underdeveloped populations. They are under-developed partly because they are nations with low local labor prices. This is a problem for the world economic system to solve. They are also under-developed because part of the labor force has low knowledge. This is a problem for the nation itself.

International competition for energy resources is clearly a factor in a nation's productivity. Natural resources are ultimately a zero sum game[6]. The Bretton Woods Conference assumed that individual productivity is the underlying source of national wealth. Nevertheless resources are an obvious aspect of productivity. As population increases, the global share of natural resources per person decreases. What a nation does with its share of resources is for that nation to decide. It is increasingly harder for an individual, region, or nation to be self sufficient. As people grow less independent, productivity is more dependent on knowledge, versatility, capital, and trade. Nations, regions, and individuals must each find a balance between consumption, productivity, and resources.

National populations have historically grown toward the limit the natural resources will support. The native population of Antarctica is zero. The native population of the Arctic is slightly higher. The United States is uniquely blessed with low population and great resources.

The relation between population and natural resources is not constant. Although agriculture began almost 10 millennia ago hunter gatherer cultures still exist. Manufacturing accelerated to prominence in 3 centuries. The changes now occur within a single generation. History is the guide to the present. It is an increasingly less reliable guide to the future.

Transportation requires energy that is portable. Petroleum demand for transportation increases faster than the growth of the supply. This increases the cost of petroleum. Transportation systems look for cheaper supplies of power. Some look toward electric power as an alternative to petroleum.

Some American automobile manufacturers believe that the ultimate future energy system for ground transportation is electric power generated by hydrogen fuel cells. The logic may seem flawed. The reaction of steam with natural gas now produces most of the industrial hydrogen. Natural gas combustion produces more heat than the hydrogen can produce. It would produce the same greenhouse gas emission either way.

An alternative is to produce the hydrogen by electrolysis. The theoretical efficiency of electrolysis is high, and the reaction is reversible. Skipping intermediate strategies for improving the efficiency of petroleum makes sense if it is assumed that the remaining era of inexpensive fossil fuel will be too short to warrant an intermediate technology. Agricultural sources of fuel cannot meet the enormous demand without an impact on the use of cropland for food production.

Whether electrolytic hydrogen makes sense depends on the cost of future electric power in relation to other sources and technology that makes hydrogen safe and easy to transport. Hydrogen production might have fundamental advantages to the electric power distribution system. If it could use off-peak demand power that otherwise has low off-peak value this would enable power plants to operate at maximum utilization. Electric power could produce the required quantities of hydrogen whenever it is available. This would give low reliability sources, such as solar and wind power, a more competitive niche. This still does not identify the source of reliable peak demand.

Notes and references for Chapter 2

1. A summary of developments leading to modern electric power generators is given in a paper by W. James King, *The Development of Electrical Technology in the 19th Century*, U.S. National Museum Bulletin, Smithsonian Institution, 1963

2. Edison is renowned for his trial and error approach to technical development.

Having little formal education, he developed a common sense approach to finding out what works and doggedly pursued his objectives. Rutgers University has published a massive four-volume compilation of Edison's papers. Johns Hopkins University Press, 1998 See also Robert J. Conot, *A Streak of Luck*, Simon & Schuster, New York, 1979

3. A continuous or dc current through a coil of wire creates a magnetic field on the axis of the coil proportional to the voltage across the coil. The magnetic field strength is proportional to the current, the number of turns in the coil, and the magnetic susceptibility of the core. Changing the current in one coil induces a current change in a second coil surrounding the same magnetic field at a voltage proportional to the strength of the magnetic field and the number of turns in the coil. The electric energy exchanged between the coils is proportional to the change in magnetic energy of the core. A transformer is a pair of coils surrounding a core. The voltage polarity alternates as a sine wave. The voltage ratio is equal to the turns-ratio. The current ratio is inversely proportional to the turns-ratio.

4. Nikola Tesla, a Serbian born in Croatia, received some education at the university in Graz, Austria, but was largely self-taught. Tesla was an entrepreneur experimenter, like Edison, but his experiments depended more on basic principles and less on trial and error. Although he never authored a paper in a scientific journal, scientific societies of England, France, and Germany gave him honors for his lectures. Tesla's normally introverted style contrasted with his flamboyant demonstrations of electrical effects. With nothing else to support his work, a large part of his life had to be devoted to raising money. The grand scale of his vision required a great deal of support. He sometimes appeared to be more showman than scientist. Seifert, Marc J., *Wizard: The life and times of Nikola Tesla*, Carol Publications, Secaucus, N.J. 1996; Margaret Cheney, *Tesla, Man Out of Time*, Delta Trade Paperbacks, New York, 1998

5. The nations are grouped by geography and cultural similarity. Europe includes nations with a potentially interconnected market. The Middle East is bounded by Morocco in the west and Afghanistan and Pakistan in the East. The Indian sub-continent includes Myanmar (Burma). East Asia includes Mongolia in the north, Malaysia in the south and Indonesia in the east. Australia and New Zealand are small and are included with North America. Mexico is assigned to Latin America, although it is geographically part of North America.

6. Change tends to be a zero sum game in which people perceive of their relative advantage as winners or losers. Lester C.Thurow, *The Zero Sum Society*, Basic Books, Inc. 1980

Chapter 3
Fossil Fuel Supply

The fuel for most of the world's electric power is in sedimentary deposits of coal, petroleum, and natural gas. The deposits occur in a wide distribution of size and quality scattered throughout the outer few kilometers of the Earth's crust. This includes the crust below the ocean surface[1]. Demand for fuel accelerated during the past century. There is promise of even greater demand. To understand the nature of the supply, the limits of the mineral fuel era, and how it will change it is necessary to know the history and properties of coal, petroleum, natural gas, and uranium deposits. It depends on properties all mineral deposits share in common starting with formation of the Earth.

The Origin of Mineral Deposits

The sun formed 4.55 billion years ago as a second or third generation star[2]. It requires at least one and probably two previous super nova cycles to account for the abundance of heavy elements like uranium[3]. The inner planets accumulated most of their mass in a hundred million years. The temperature of the Earth was above 20,000 C. The Earth began with little or no gaseous atmosphere since gases had enough kinetic energy to escape.

The inner planets, Mercury, Mars, Earth, and Venus are small. They have high densities and high concentrations of heavy elements. The outer planets, Jupiter, Neptune, Saturn, and Uranus are much larger. They have much lower densities of gases that surround a relatively small core of heavier elements.

The Earth's core has three solution phases[4]. The center is a solution of metallic iron and iron oxides. It is a solid at that high pressure. The Earth's rotation magnetizes it. This generates the magnetic field that forms the *Van Allen belt* of trapped electrons and protons of the solar wind at a 200 km altitude and prevents them from stripping the Earth of its atmosphere.

Table 3.1 lists the major components of the Earth in decreasing order of mass in petatons (10^{15} metric tons, Pt). Fossil fuels are a small part of the carbon distributed among these components.

Table 3.1 Components of the Earth's mass

Earth total mass[5]	5,980.000 Pt
Earth core[5]	1,910,000 Pt
Earth mantle[5]	4,050,000 Pt
Lithosphere (crust)[6]	23,600 Pt
Hydrosphere (oceans)[6]	1,370 Pt
Atmosphere[7]	5.3 Pt
Biosphere[8]	2190 Gt
Living matter in biosphere[8]	610 Gt

The mantle is a semi-liquid phase surrounding the core. It is a solution of iron oxides with a ferric/ferrous ratio slightly less than one. The Earth's core thus has a reducing environment. The normal form of nitrogen and carbon is hydrides, ammonia and methane, rather than oxides.

A less dense outer phase is a solution of silica and metallic oxides. Convection constantly stirs this phase. This causes *tectonic plates* to spread from rifts in the ocean bottom. They are the material that forms the Earth's outer crust.

The Earth's crust or lithosphere is basalt rock formed as the mantle solidified over 4 billion years ago[9]. Its thickness averages 35 kilometers over the continents and 4.5 kilometers beneath the oceans. A thin layer insulates the surface from the high temperature core[10]. The surface temperature is a balance of thermal conductivity from the core, radiation from the sun, and thermal radiation into space.

Sheets of basalt spread from the mid-ocean rifts and constantly replace the bottom of the lithosphere. Where they meet the continental tectonic plates the denser layer pushes underneath. This *subduction* returns them to the mantle in about a 100 million year cycle. The less dense layer creates mountain ranges and other topographic features. Basalt rock of the lithosphere has roughly the same elemental composition as the much more massive mantle.

Oceans cover two thirds of the Earth's surface at an average depth of 3794 meters. This is a small fraction of the water in the Earth's core. *Fumerals* of hot water and dissolved gases flowing from mid-ocean rifts, hot springs, and gases released by volcanic eruptions are the major sources of ocean water. The higher density crust rocks cover the other third of the surface. They have an average altitude of 875 meters above sea level. This balances the displacement of the oceans below sea level.

The mass of the oceans, continents, and atmosphere is less than 1 percent of the mass of the mantle. The large ratio leaves the mantle composition constant over geologic time. Processes of the upper atmosphere preferentially deplete light elements. Rainfall and rivers cause land erosion. The oldest geological sediments are deposits of basalt shale in the ocean.

The atmosphere is the cumulative quantities of substances in volcanic gases released from the mantle through fissures and volcanoes. It forms continuously. Table 3.2 compares the composition of the major volcanic gases with the composition of the present atmosphere[11, 12]. Nitrogen gas is only 1.3% of the volcanic gases. Since it remains mostly in the form of N_2 gas its accumulation over geologic history makes it the dominant atmospheric gas. The cumulative mass of other elements that originate as volcanic gases can be estimated from their mole percent relative to nitrogen in volcanic gases[13].

Table 3.2 Comparison of the mole percent of major volcanic gas components with the present atmosphere (percent by volume)

	Volcanic	present	cumulative mass
H_2O	79.3	0 - 4	3.60×10^{19} kg
CO_2	11.6	0.038	1.29×10^{19} kg
SO_2	6.5		1.05×10^{19} kg
N_2	1.3	78.08	9.17×10^{17} kg
CO	0.4	-	2.82×10^{17} kg
Cl_2	0.05	-	8.95×10^{16} kg
H_2	0.6	-	
Ar	0.04	0.934	
O_2	-	20.95	2.81×10^{17} kg

Carbonate rock deposits contain 99.5% of the carbon outside of the mantle. The carbon originates as atmospheric CO_2. Its solubility in the ocean depends primarily on the partial pressure of CO_2 in the atmosphere, the concentration of Ca^{++} and Mg^{++} ions, and the *pH of the ocean*, its acidity or alkalinity. Ca^{++} and Mg^{++} ions are weak bases which control the pH. The present ocean has a slightly alkaline pH = 8.1. At this pH atmospheric CO_2 dissolves in the oceans as bicarbonate ions, HCO_3^-.

Ca, Mg, Na, and K are elements of the primary crust rock that are soluble. They dissolve from rocks and flow to the ocean as Na^+, K^+, Ca^{++}, and Mg^{++} ions. They cause the high salt concentration in ocean water. Large deposits of NaCl salts formed from oceans that evaporated when they became isolated from a water source.

The major carbonate rocks are deposits of the calcareous remains of shelled sea animals. Crystalline carbonate minerals deposits form mixtures of calcium and magnesium carbonates with hydrous magnesium aluminum silicates in equilibrium mixtures like the following[14].

$$Mg_5Al_2Si_3O_{10}(OH)_8 + 5\ CaCO_3 + 5\ CO_2 =$$
Chlorite   calcite

$$5\ CaMg(CO_3)_2 + Al_2Si_2O_5(OH)_4 + SiO_2 + 2\ H_2O$$
dolomite kaolinite

According to Table 3.2 the total abundance of carbon is 12 mole percent of all volcanic gases. The combined abundance of Ca and Mg is sufficient to deposit most of the original CO_2 as carbonate rocks. The oceans contain most of the remainder in the form of HCO_3^- ions in equilibrium with Ca^{++} and Mg^{++} ions. The temperature, pH, and HCO_3^- ion concentration control the residual atmospheric pressure of CO_2 at equilibrium. The massive size of the ocean does not guarantee equilibrium concentrations.

Table 3.3 outlines the geological history of the Earth. Successive layers of sediment are the primary record of geological history. Stratigraphy is the science that analyzes it. For the first billion years the history was mainly inorganic mineralogy. Significant organic biological history did not begin until the Earth had an oxygen atmosphere.

Table 3.3 Geological eras and rough dates in millions of years ago identified by fossils and minerals starting with the formation of the Earth at the bottom

Cenozoic Era
 Quaternary Period
    Pleistocene   5.3   ape fossils
 Tertiary Period
    Pliocene
    Miocene   25   grasses
    Oligocene   37   cats, dogs, pigs, bears
    Eocene   55   hoofed mammals, rodents, whales
    Paleocene   67   primates
Mesozoic Era
    Cretaceous   138   demise of dinosaurs
    Jurassic   208   birds
    Triassic   245   dinosaurs
Paleozoic Era
    Permian   290   flowers, insect pollination
    Carboniferous 360   conifers
    Devonian   410   trees, vertebrates on land
    Silurian   435   spore bearing plants
    Ordovician   520   land animals
    Cambrian   570   vertebrates
Proterozoic Era
              620   brains
              670   jellyfish
            1000   sexual reproduction
            1300   land plants
Archean Era   3500   photosynthetic bacteria
Hadean Era
            3600   oceans, sediments, bacteria
            4000   basalt rocks
            4550   earth formed

Evolution of an Oxygen Atmosphere[15]

The earliest fossil evidence for life is bacteria in 3.6-billion year old shale[16]. There is little direct knowledge of the world during the 400 million years after the Earth's crust first formed. Rivers formed deposits of volcanic ash and mud that became shale and slate. Numerous experiments have fueled speculation over how life started.[17] To form a single-cell animal complete with a nucleus, a membrane, and other cell components in the mud environment in which this cell was found requires a convergence of unusual circumstances.

The massive concentration of ferric and ferrous iron oxides in the mantle determines the oxygen and oxidation-reduction balance of the Earth. The traces of H_2, CH_4, NH_3, and CO and absence of O_2 in volcanic gases are consistent with the Fe_2O_3/FeO ratio slightly less than one, a reducing environment.

The Earth's crust is a barrier that allows the matter the Earth contains to have a different oxidation state outside and inside the crust. Since the mass of oxygen in the iron oxides of the mantle is 4 million times the mass of oxygen in the present atmosphere the environment outside the crust changed to its present oxidizing state with little effect on the mantle. The present abundance of O_2, H_2O, CO_2 and carbonate rocks shows that the Earth's surface has in fact changed from a reducing to an oxidizing environment.

Organic reactions with small free energy differences can produce a vast array of products from similar starting materials. However biological molecules have much higher free energy than CO_2, the dominant source of carbon.

Photosynthesis provides the energy to convert CO_2 to bio-chemical molecules. Regardless of how life began, *chlorophyll* is the catalyst that allows the cumulative energy of solar photons to produce a nutrient molecule. Biological nutrients such as glucose require the energy of several solar photons.

$$6\ CO_2 + 6\ H_2O + >12\ h\nu \rightarrow C_6(H_2O)_6 + 6O_2$$

Cells are the basic unit of a living organism. Unlike the unit cell of an inorganic crystal they contain a collection of many chemically independent components. The DNA nucleus of a cell is the template that allows a cell to replicate the nucleus. This is the observable basis for study. What causes the initial DNA to form remains speculative. The same mud as the earliest biological molecules is a catalyst in petroleum refining.

The big factor in biology's favor is time. Once life exists *natural selection* can improve it. In a suitable environment a bacterium divides every 20 minutes. In a 12-hour period this gives 2^{36} or 70 billion opportunities for a mutation. The next 12 hours gives 70 billion opportunities to test whether the mutation propagates or dies out. Even under less than ideal conditions the 150 billion day period gave life uncountable opportunities to evolve. How a membrane that encloses a DNA nucleus with a group of symbiotic, but independent entities evolved is remarkable, but not entirely mysterious.

Blue-green algae appeared 100 million years after the first bacteria. The ocean shielded it from ultraviolet radiation damage. The algae absorb solar energy and CO_2 to produce nutrients and oxygen. Land plants and animals could not evolve until a more favorable environment began to form about 3 billion years ago.

Ultraviolet radiation from the sun changes the composition of the upper atmosphere as successive layers absorb it. The radiant power depends directly on the number of photons/sec and their energy. It decreases sharply with increasing photon energy or decreasing wavelength.

At the top of the atmosphere the highest energy photons dissociate water vapor and other molecules that contain hydrogen atoms. The hydrogen atoms have enough kinetic energy to escape the Earth. Ultraviolet radiation depletes water vapor and other hydrogen containing gases at the top of the atmosphere. Hydrogen depletion shifted the equilibrium oxidation state outside the mantle from the reduced states, H_2, CH_4, NH_3, and CO, to oxidized states.

The Earth continuously loses hydrogen by photolysis of molecules that contain hydrogen from a layer at the top of the atmosphere. To reach the top of the atmosphere additional hydrogen containing molecules must diffuse through other gases, mostly nitrogen and oxygen. This is a very slow process. After the initial period the rate of hydrogen depletion decreased.

Oxygen absorbs the highest energy photons that penetrate the layer of molecules containing hydrogen. It forms oxygen atoms. Oxygen molecules O_2 greatly outnumber the oxygen atoms. The atoms combine with oxygen O_2 to form ozone O_3. As oxygen accumulates, ultraviolet photolysis creates a layer of ozone. Ultraviolet radiation destroys much of the ozone during each day. It returns each night.

$$O_2 + h\nu = 2\,O$$

$$O_2 + O = O_3$$

$$O_3 + h\nu = O_2 + O$$

Ozone inhibits plant life. However when ozone and oxygen block ultraviolet radiation at the top of atmosphere the nitrogen and oxygen at lower altitude become a diffusion barrier that protects plants. Until the oxygen reached 1 percent of its present concentration ultraviolet radiation reached the ground and prevented plant life. As oxygen accumulated the ozone layer rose to higher altitude. The lower atmosphere became hospitable to plant life on the Earth's surface and life emerged from the ocean.

While these changes in the planet were taking place air and water flow deposited successive layers of sediments. The geological record of mineral changes covers the entire planet. This created a record of subsequent alterations by geologic history that is data for stratigraphers to analyze.

Economics of Mineral Deposits[18]

Dead plants and animals decay to form the coal, petroleum, and natural gas deposits we now consume as fuel. The deposits are in sediments that cover much of the lithosphere and ocean bottom. They occur in thousands of geological layers in widely dispersed pockets in particular layers from the surface to a depth up to 3 kilometers.

The fraction of deposits large enough and concentrated enough to be useful is small[19]. The quantity of fuel at a single site or set of closely related sites is the most relevant property to a fuel producer. The value of the deposit increases with size, quality, and accessibility. It decreases with the processing required to extract useful value, and the cost to remedy environmental quality of the fuel or the damage in mining it.

The relevant property of all deposits of a given mineral is their size distribution. The fraction of deposits that remains undiscovered is unknowable. The characteristics of the known deposits give statistical clues that narrow the uncertainty about future availability and cost. These clues plus knowledge of geology determine whether exploration to find new deposits is likely to be profitable. This proceeds in three steps.

Discovery means prospecting to find new deposits. Knowledge of the geology of known deposits starts by understanding how geologic processes create the mineral, form deposits of it, and position it in a particular layer. The sequence of layers dates the formation process in geological time. The present depth and orientation of the layers depends on subsequent events. These include continental drift, different climates, changes in temperature and pressure as the land rises and falls under the weight of glaciers, and tectonic plate movements.

Seismic prospecting reveals the orientation and structure of deep layers[20]. This may identify hundreds of pockets that might hold a deposit. It does not determine whether a pocket is a fuel deposit. This requires a core drill to the depth and location of the deposit. Only real samples reveal the fuel quality. The rare test hole that discovers a deposit reveals its quality but not its quantity.

Development expands the discovery area to determine the extent of the deposit. Investment decisions rely on this information about the quantity, quality, and cost of whatever fuel is available. Each original discovery requires at least four additional test holes to determine the direction of the field. Large deposits have greater yield per test hole. Kansas has the most one-hole oil fields of any oil producing area in the world. The decision whether to produce a mineral depends on the mineral. Natural gas is the easiest production decision. For the most part it simply flows from the test holes. Production depends on whether pipeline facilities are available to receive it.

Petroleum generally does not occur as a liquid pool but as a liquid which seeps through sand or porous rock. Whether the flow rate is adequate depends on the viscosity of the oil as well as the porosity of the oil field. All petroleum is not the same. Different refinery methods must match the petroleum qualities to the application.

Coal, like other solid minerals, requires machines that remove the mineral from the mine and through whatever processing is necessary. The traditional practice is simply to burn coal in a furnace. However the quantity of pollutants this generates is no longer tolerable. It is increasingly necessary to remove pollutants from most grades of coal by refining.

Uranium requires extensive refining. Facilities at the mine recover pure natural uranium oxide from low grade ore. Very special facilities then enrich the isotopic components of uranium that produce nuclear fission.

Production may require multi-billion dollar spending decisions by both electric power producers and mine owners. The primary factors are the type of fuel and size of the deposit. Electric power producers require sources large enough to be a dependable supply over the lifetime of a large and expensive power plant. The fuel producer requires a deposit large enough to justify the expensive installations necessary for robust production with large economies of scale.

Proven reserves are the data the development process provides. The *reserves-to-production ratio* is the total quantity in the deposit divided by the current annual production rate. It has a value in years. It is how long the mineral in that particular deposit will last at the current rate of production.

This is misleading. A deposit that is fully developed but not now economical to produce has zero production and an infinite reserves-to-production ratio. The economics of fuel production does not depend on how much fuel exists, but on how much is economical to produce. This depends on the specific properties of each individual oil field or mine.

Since a robust market in fossil fuels exists the demands of the market tend to trump all other considerations. Economics guarantees that the first deposits to be used are the largest and most valuable[21]. This has a number of consequences that are not intuitively obvious. Deposits with the lowest price are the first to be used because the cost of the large capital investment must either yield a return or be written off as a loss.

The largest deposits are typically the first consumed and cheapest to produce. The scope of prospecting increases the probability of finding large high quality deposits first. The quality of deposits decreases geometrically. For entities that can be ranked in order of decreasing size *the size times its rank in the distribution is often a constant*[22]. According to this rule the dozen largest oil deposits contain half of the oil to be found. The deposits yet to be discovered are governed by the decreasing odds that the next discovery will be among the largest in history.

The size and quality of undiscovered reserves tends to decline continuously. Major users need to assure large quantities of fuel from large deposits of the same quality over long periods of time. As long as energy is cheap people invent new ways to use it. These are conditions that end reliance on a declining resource.

The oil geologist Maynard King Hubbert found 30 years ago that a year-by-year graph of the size of new oil discoveries was approaching what appeared to be the peak of a Gaussian shaped curve[23]. Subsequent history confirms that behavior.

Petroleum and Natural Gas Formation

The discussion of fossil fuel formation starts with petroleum because it has the longest history. Although petroleum is no longer a competitive fuel for electric power, its increasing scarcity and cost has implications for electric power. Petroleum originated primarily as microscopic fossil sea animals. The history starts with the history of the oceans. The process from sea animals to petroleum requires millions of years.

Petroleum geology starts at river basins where mud and weathered rocks from land erosion form the mud clay precursors of shale deposits. These become petroleum source rocks. Clay sediments are not stationary. They migrate with the wave motions and tides, filling low levels and geological faults. The mud collects organic matter from various sources, chiefly dead marine organisms.

Liquid petroleum consists primarily of aliphatic hydrocarbon chains of up to about 10 carbon atoms. The hydrogen-to-carbon mole ratio must be at least 1.5 for liquid petroleum. Plant carbohydrates have ratios less than 1 and do not form petroleum but may produce natural gas. The lipids and proteins of dead marine organisms have chains that are long and complex. They can ultimately yield H/C ratios greater than 1.5.

The clay mud of river deltas is a shallow ocean environment with enough sunlight to provide energy for organism to grow. Petroleum formation starts when the sea animals die and lie in an oxygen free underwater clay mud environment. A sequence of steps transforms the dead organic matter to petroleum.

Biochemical decomposition is the result of reactions of enzymes, bacteria, and other microorganisms. These reactions reduce the lipids and proteins of sea animal remains to high molecular weight solids known as *kerogens*. The biochemical reactions take place at ambient temperatures before geologic processes bury the material to a depth at which there is a significant temperature increase. The kerogens are not necessarily in coherent deposits. They may be scattered throughout the mud deposits that capture them.

Physical processes that continue the transformation require heat and pressure. All petroleum forms at a depth between 2500 and 5000 meters. Here temperatures in the 75 to 150°C window are high enough to break the bonds in large molecules without producing only natural gas. Temperature and pressure over time transform the kerogens to aliphatic hydrocarbons, a process similar to the catalytic cracking reactions in oil refineries. The maturity of the final petroleum depends on the extent to which the thermal reactions remove oxygen, nitrogen, and sulfur. The mud and silt sources eventually become clay shale and sandstone *source rock*. Gas formation creates high pressure that drives the liquid petroleum out of the source rock.

Water flow through rock masses may not seem possible. On a geologic time scale it is not only possible but is a major mechanism that forms mineral deposits. The interstices in permeable rocks are hydrophilic. Water lubricates the flow of oil, particularly at temperatures and pressures that increase the solubility of oil. The oil-water mixture flows through voids and permeable sandstones.

Petroleum flow is substantial once it becomes a gas or liquid. Petroleum can form only in the temperature window at a depth of 2500 to 5000 feet. It seldom remains at the deep level needed to form it. To form a deposit the petroleum must reach a *trap rock* formation. The trap rock formation may be far from the original source rocks. It may be at or near the surface, as in some Pennsylvania and Saudi Arabian fields. An oil basin might be oil formation geology.

A petroleum reservoir is typically a layer of permeable sandstone between a dome of impermeable limestone or salt and an impermeable lower layer at depths of 1000-3000 meters. Typically a sandstone layer containing the oil deposit has water saturated with oil on the bottom, oil in a middle layer, and gas on top. For a time the gas pressure may force the liquid out of the well.

Ultimately flow through the porous rock to the wellhead limits the production rate of a well. The flow is usually slow and intermittent. Most of the oil is unrecoverable. Secondary recovery methods can produce some additional oil.

Fig. 3.1 compares the oil production and consumption of nations with the largest production and/or consumption in GW-yrs of heat energy. The quantities exchanged in trade are shaded portions of each bar. More petroleum is traded than any other commodity. The primary uses of oil are transportation and residential heating. Demand for these purposes now makes it too expensive for new electric power plants. Its increasing cost makes it irrelevant to electric power. When all else fails people rely on electric power as a generic energy source that can solve all problems.

Producer nations are nations that produce more oil than they consume. The length of the bar represents the total production by these nations. The un-shaded part is the consumption. The shaded part is the portion they export to other nations.

Consumer nations are nations that consume more oil than they produce. The length of the bar represents the total consumption by these nations. The un-shaded part is the production. The shaded part is the net imports from other nations. Nations rely on trade to balance their production and consumption of petroleum more than for any other commodity. A robust market establishes prices according to the properties of the oil. The trade is bilateral mainly in the sense that consumer nations and oil refining companies have an interest in providing capital support for oil producers that are likely to provide a reliable supply.

The number following the bar for producer nations is the *reserves-to-production ratio* in years. The largest deposits ever discovered still supply much of the oil. They deliver it at an enormous rate. Deposits don't run out of oil. The rate at which wells can deliver oil decreases as the oil sand slows the flow to the well head.. Typically pumping can recover only about 30% of the oil. The remaining reserves require secondary recovery methods using steam, detergents, solvents, and high pressure.

When the demand for a mineral is strong, the rate of discovery of new deposits is what limits production. The cumulative recoverable oil in new discoveries as a function of the number of discoveries might logically have a bell shape. If so, petroleum reached its maximum in the 1970s[24].

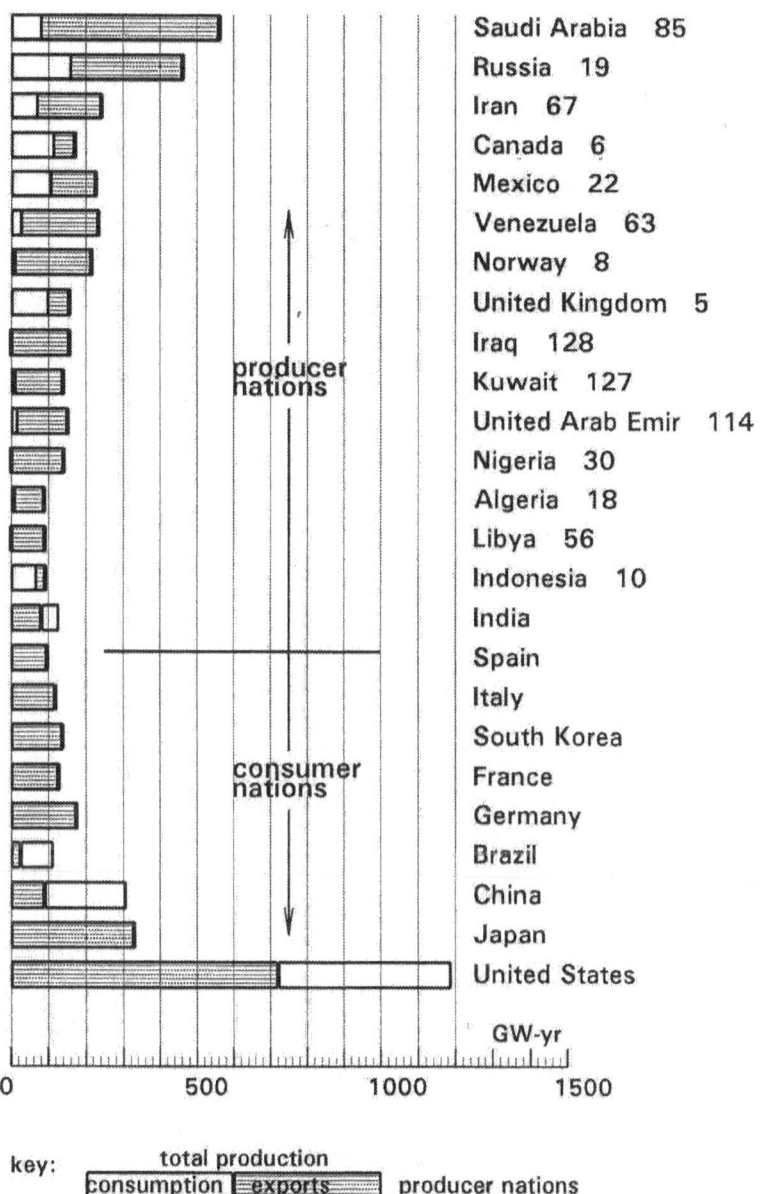

Saudi Arabia   85
Russia   19
Iran   67
Canada   6
Mexico   22
Venezuela   63
Norway   8
United Kingdom   5
Iraq   128
Kuwait   127
United Arab Emir   114
Nigeria   30
Algeria   18
Libya   56
Indonesia   10
India
Spain
Italy
South Korea
France
Germany
Brazil
China
Japan
United States

producer
nations

consumer
nations

GW-yr

0   500   1000   1500

key: total production
[consumption | exports]   producer nations
          total consumption
[imports | production]   consumer nations

Fig. 3.1 Oil producers and consumers
Data source: British Petroleum Company 2004

Natural gas deposits are much more prevalent than oil deposits. Pipeline distribution networks analogous to the electric power grid connect the grid of gas wells directly to a grid of a large fraction of consumers.

Fig. 3.2 compares the natural gas production and consumption of nations with the largest production, consumption, exports, imports, and proven reserves. The natural gas consumption is for both electrical power generation and residential and industrial heating. The length of each bar and its shading distinguish producing and consuming nations as before.

Producer nations produce more natural gas than they consume. The length of the bar represents the total production by these nations. The un-shaded part is the consumption. The shaded part is the portion they export to other nations.

Consumer nations consume more natural gas than they produce. The length of the bar represents the total consumption by these nations. The un-shaded part is the production. The shaded part is the net imports from other nations.

At ordinary temperature and moderately elevated pressure the higher molecular weight propane fraction of natural gas is a liquid. The demand for gas by more isolated consumers creates a market for liquefied natural gas. The high energy density of the liquid makes it practical to refrigerate liquefied gas and ship it in international trade in very large volume containers.

Natural gas burns at a high enough temperature to drive a gas turbine engine. This provides much greater flexibility to start, stop, and change power level than a coal fired steam engine. Although the fuel energy is significantly more expensive it can provide reliable peaking power that coal cannot from plants that are smaller and cheaper to build.

The reserves of natural gas associated with coal and oil field geology might be only a fraction of the methane gas that must exist. At high pressures or ice temperatures methane forms a hydrate with water that can exist as a liquid. A 1997 survey estimated that the potential methane as a hydrate in deep oceans could far exceed the fossil fuel carbon.[25]

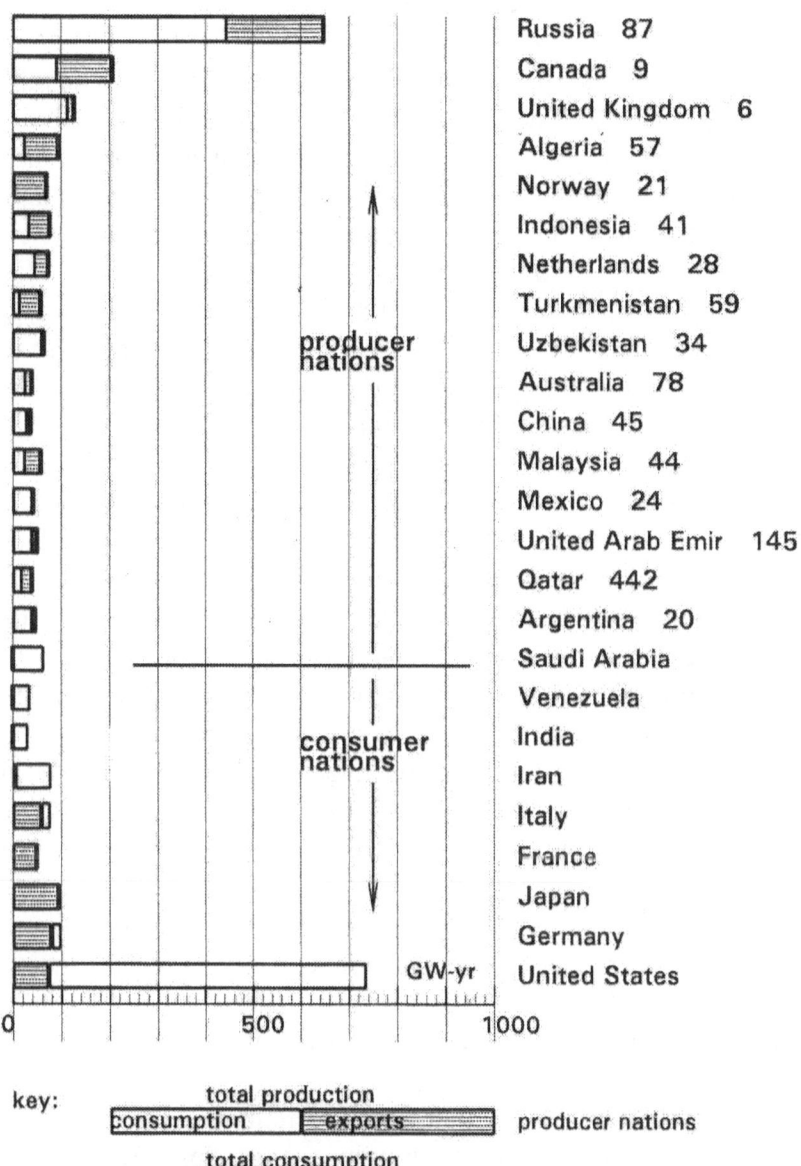

Russia 87
Canada 9
United Kingdom 6
Algeria 57
Norway 21
Indonesia 41
Netherlands 28
Turkmenistan 59
Uzbekistan 34
Australia 78
China 45
Malaysia 44
Mexico 24
United Arab Emir 145
Qatar 442
Argentina 20
Saudi Arabia
Venezuela
India
Iran
Italy
France
Japan
Germany
United States

producer
nations

consumer
nations

GW-yr

0   500   1000

key: total production
            consumption   exports   producer nations
            total consumption
            imports   production   consumer nations

Fig. 3.2 Natural gas producers and consumers
Data source: British Petroleum Company 2004

Coal Formation and Consumption[26]

Land plants emerged during the Cambrian age 570 million years ago but were not abundant until the coniferous period 200 million years later. During that time an oxygen atmosphere formed and pushed the ozone layer to an altitude where it shields plant life. The North American plate was probably a tropical climate more favorable to rapid growth.

Coal comes from plant growth in swamp land. Deposits often contain giant fern fossils with woody stems up to 70 feet high. To form peat biomass the dead trees avoid oxidation by remaining covered by water as they are buried by other dead trees. The water must remain deep enough to cover the growing layer of peat yet shallow enough for lush forest growth that is rapid in comparison to other sedimentation. As the biomass depth accumulates under shallow water the land must subside at the rate biomass is deposited.

Coal deposits currently being mined in the U.S. surface mines are 0.2-to-10 m thick. The minimum thickness for underground mines is greater. The peat deposit must be 3 to 5 times the thickness of the eventual coal deposit. Peat biomass is converted to coal in steps. The peat bed must remain intact throughout the geological events and processes that convert it to a useful coal deposit. It is ultimately buried by sediment where it may be subjected to tectonic action.

Biochemical reactions start coal formation with an attack on plant carbohydrates by algae, bacteria, and enzymes. This increases the relative carbon content and eliminates water. The biochemical phase ends when the carbon content is about 60%. The peat is then *lignite coal*.

Physical processes that continue require some combination of high temperature and/or pressure over time. Reducing the carbohydrate water content of the coal increases the heat energy. Different combinations of high pressure, high temperature, and time produce similar results. As the carbon content increases beyond 60% to 70%, 80%, and 90% the coal progresses beyond *lignite to sub-bituminous*, *bituminous*, and *anthracite* coal, respectively.

Anthracite coal is formed in deeper mines under more severe conditions. Distilling the volatile molecules from coal produces *coke*, relatively pure carbon. The coal tar distillate contains aromatic hydrocarbons used as feedstock for the chemical industry. Anthracite used for coke is known as *metallurgical coal*. The high heat of combustion of coke can reduce metal oxide ore to pure metal or produce lime, CaO, from limestone, $CaCO_3$.

The heat content of coal determines the price. Except for environmental damage lower cost grades are technically suitable for electric power. A significant fraction of the lignite, sub-bituminous, and a few bituminous deposits occur in sites that can be mined from the surface by stripping off the overburden. The depth and size of the deposit determines how much overburden is economical to remove.

The size of economical deposits ranges from about 10,000 to well over a million tons per year. Larger scale deposits justify larger scale equipment by economies of scale. Larger scale deposits give more predictable supply over the life of a power plant. The size of the deposit determines whether it is worth removing the overburden to create a surface mine. The three major U.S. coal basins illustrate how different classes of coal and mines complicate the interpretation of reserves-to-production ratios. The basins were formed in three different geological eras.

Coal formation ends when the peat bog is flooded, the vegetation is drowned, and sedimentation buries the deposit. Bogs in the interior basin of the U.S. were flooded by ocean water. This produced limestone and gypsum deposits with high sulfur content. Bogs flooded by fresh water are covered by shale deposits with low sulfur content.

Appalachian deposits in West Virginia, Pennsylvania, Kentucky, and Ohio contain most of the high quality, low sulfur anthracite metallurgical coal. The large underground mines use continuous high capacity mining technology[27], but the scale is not comparable to mines that now supply electric power plants.

Industries use coal for process heat in addition to electricity. The metallurgy industries use high grade anthracite coal to produce high temperature. It comes from deep underground mines. This is generally higher quality than the electric power industries require. As environmental quality of available coal deteriorates it is more realistic to refine low quality coal than depend on high quality.

Fig. 3.3 compares the coal production and consumption of nations with the largest production and/or consumption. The quantities are in GW-yrs of heat energy[28]. Electric power generation plants consume more than half the total. One GW-yr of electric power is equivalent to about 1.8 million tons of coal. Coal and uranium are at present the only fuel sources with the capacity to supply this level of energy. A world demand of 5000 GW is equivalent to 25 million tons of coal per day.

Producer nations are nations that produce more coal than they consume. The length of the bar represents the total production by these nations. The un-shaded part is the consumption. The shaded part is the portion they export to other nations.

Consumer nations are nations that consume more coal than they produce. The length of the bar represents the total consumption by these nations. The un-shaded part is the production. The shaded part is the net imports from other nations.

The number after the name of the producing-nations is the reserves-to-production ratio of the developed mine. The previous discussion described the cautions about interpreting this as years of guaranteed production.

Nations produce most of the coal they consume. Coal is a high density energy source that is easy to transport. Nevertheless export-import trade accounts for a relatively small fraction of world coal production. The United States and China are by far the dominant coal producing nations. Russia has significant under-utilized capacity. China and India will almost certainly exceed the U.S. demand for electrical energy. Japan and South Korea are the major coal importer nations. They rely on imports for metallurgy. All of the coal deficient nations are increasing their reliance on nuclear power as the alternative to coal for electrical power.

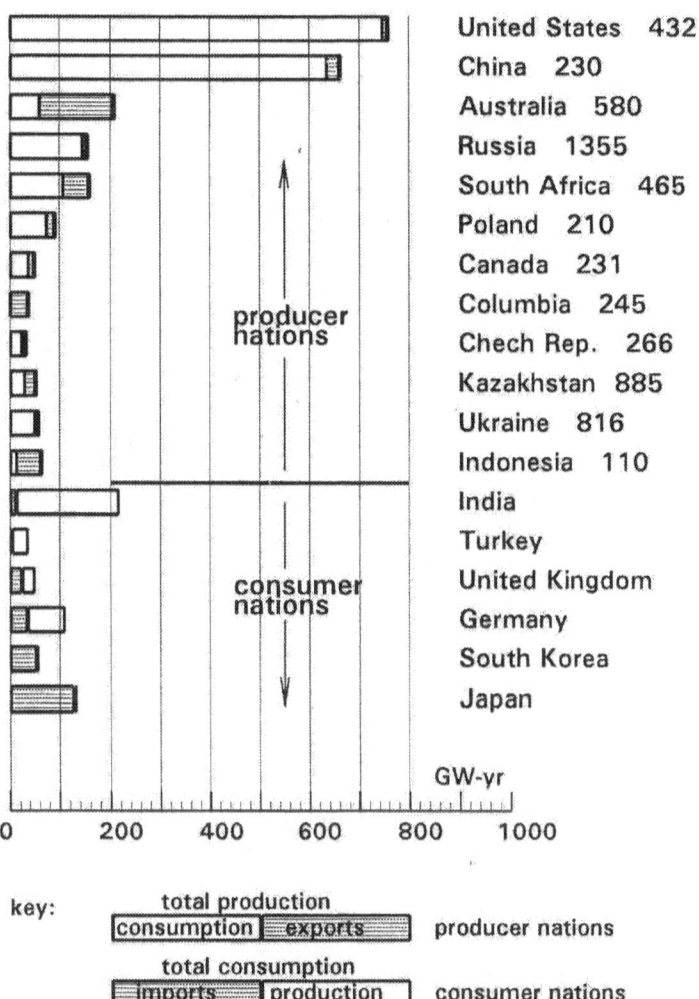

United States   432
China   230
Australia   580
Russia   1355
South Africa   465
Poland   210
Canada   231
Columbia   245
Chech Rep.   266
Kazakhstan   885
Ukraine   816
Indonesia   110
India
Turkey
United Kingdom
Germany
South Korea
Japan

producer nations

consumer nations

GW-yr

0   200   400   600   800   1000

key:

total production

| consumption | exports | producer nations

total consumption

| imports | production | consumer nations

Fig. 3.3 Coal producers and consumers

Data source: British Petroleum Company 2004

Table 3.4 lists the reserves of different classes of coal in the three United States basins in millions of metric tons.

Table 3.4 Current consumption of U.S. Coal Reserves

	production Mt	operating mine reserves, Mt	r/p	total reserves Mt	r/p
Western basin					
Underground	48	817	16	145029	3085
Surface	**463**	**11231**	**24**	**95603**	**206**
Total	511	12048	23	241525	472
Interior basin					
Underground	54	1094	20	119586	2214
Surface	90	1307	14	44595	495
Total	144	2491	17	164490	1142
Appalachian basin					
Underground	272	2942	10	74668	274
Surface	147	859	5	22889	155
Total	419	3801	9	97557	233
U.S. Total					
Underground	374	4853	12	339282	907
Surface	700	13486	19	163378	233
Total	1074	18339	17	502660	468

Of 5000 developed coal deposits in the U.S. about 1400 mines currently produce coal. The major distinction in the reserves-to-production ratios is between mines that are operating and those that are not. The reserve ratios are most significant for individual mines operating at full capacity.

The demand favors specific mines that deliver very large volumes of coal that can meet environmental quality standards with the least processing. Few if any of the active mines can continue to deliver coal at the present rate over the lifetime of a power plant. Economics dictates a trend toward mines that produce lower volume and/or lower quality in some respect.

The western basin is the largest and most recent. About 40% of its developed deposits are surface mines. Deposits in Montana, Wyoming, and Colorado produce by far the largest share of the coal for electric power. Their low sulfur lignite or sub-bituminous coal can be produced at a fourth the cost of coal from underground mines.

The interior basin is intermediate in size and age with 27% of the coal in surface mines. Coal from Illinois, Indiana, and 8 other Midwestern states has high sulfur content. The sulfur causes environmental problems beginning with sulfuric acid aerosols that produce atmospheric smog and acid rain. Sulfur content ranging from less than 1% to more than 3% is a result of how the deposit was ultimately buried.

The Appalachian basin is the oldest in age and smallest in size with 23% of the coal in surface mines.

As the active mines are exhausted, production shifts to coal of lower value and/or higher cost. Western low sulfur surface mines make up only about 15% of known U.S. coal reserves. The value of coal must decrease continuously as the deposits become smaller, lower quality, and/or less accessible.

Ultimately coal will either fail to supply adequate volume at a tolerable cost or fail to satisfy environmental requirements. High sulfur coal in the quantity required for electric power produces too much environmental damage to use directly. Removing sulfur from flue gases is expensive and inefficient. Emerging technology using powdered coal may make it possible to remove the sulfur much as it is removed in refining petroleum.

Coal is one of the few sources with the capacity to supply reliable base power. The supply of coal will never be completely exhausted. In one sense it is quite correct to say that we will never run out of coal. Each new power plant must either face a new set of more complicated, less favorable, and more expensive choices or use an alternative source of energy.

It requires 50-80 carloads per day to satisfy the demand of an electric power plant. This is the primary requirement any feasible combination of alternative sources of energy must satisfy.

Notes and references for chapter 3

1. The deepest mine in South Africa is 3600 meters. The deepest oil well in Elk City, Oklahoma, is 9,599 meters. Access to deep deposits is limited by the *geothermal step*, typically an increase of 1°C for every 25-35 meters.

2. The age of the Earth is given by the $^{87}Rb/^{87}Sr$ ratio in mass spectra of chondritic meteorites assuming that (1) these meteorites are formed from the same materials that formed the Earth and have the same heavy element composition as the Earth, (2) the radioactive ^{87}Rb is a product of the nuclear reactions of the super nova explosion that created the Earth, and (3) and the ^{87}Sr product of the ^{87}Rb-to-^{87}Sr decay (half-life of 49 billion years) did not occur before the Earth was formed.

3. Elements up to the atomic mass of iron form by successive reactions at the temperature of hydrogen fusion. The iron abundance according to the spectrum of the sun and the density of the inner planets is consistent with nuclear reactions of a previous sun before its destruction as a super nova. It may have required fusion reactions from more than one previous super nova cycle to yield elements with atomic weights much greater than iron. Geoffrey Zubay, *Origins of Life on Earth and in the Cosmos*, Academic Press, New York, 2000

4. The density and diameter of phases in the core are from reflection and absorption of seismic waves during earthquakes. The elements of the core and mantle are from the composition in chondrite meteorites.

5. The Earth's geometric parameters are refinements of measurements that start with latitude. Over the period of a year the angle of the zenith Sun from the vertical is equal to the latitude ± 23.45°, the component of the tilt of the Earth's axis with respect to the plane of the solar orbit in the North-South plane. The tilt component vanishes at each equinox. The Earth circumference is a simple proportionality of the ratio between the latitude at two points on the same longitude and the distance between them. The radius, surface area, and volume are determined from the circumference. The mass of the Earth is obtained from values for the radius, the universal gravitation constant, and the gravitational acceleration at the Earth's surface using Newton's *Law of Universal Gravitation*. The composition of the Earth's interior is from seismic reflections during earthquakes from boundaries found by proportioning the mean density (5.5 g/cc) among metallic iron of the inner core (7.9 g/cc), iron oxides of the outer core (5.1-5.7 g/cc), and basalt magma of the mantle (3-5.5 g/cc). Kenneth R. Lang, *Astrophysical Data: Planets and Stars*, Springer Verlag, New York, 1991

6. The mass of the lithosphere (outer crust) is based is on the mean density (2.7 g/cc) and the volume given by a seismic sounding map of the thickness of the Earth's crust at 1° intervals.

    The mass of the hydrosphere is the density of water times the volume given by a sonar map of the depth of the oceans and other major bodies of water. This excludes the water in the lithosphere.

7. Half of the mass of the atmosphere is below an altitude of 17.4 km (57,400 ft). Robert C. Weast, Ed., *Handbook of Chemistry and Physics*, Chemical Rubber Company, West Palm Beach, FL 2001

8. The total mass of living matter has been obtained from systematic samples of plants, animals, and their chemically intact remains. Robert H. Whittacre, *Communities and Ecosystems*, McMillan Publishing Co., New York, 1975. This is now augmented by global satellite measurements. J.S. Olson, et.al., *Major World Ecosystem Complexes Ranked by Carbon in Live Vegetation: A Database*, Oak Ridge National Laboratory, Carbon Dioxide Information Analysis Center, 2001

9. The age at which a particular rock formed from the mantle is given by its $^{40}K/^{40}Ar$ ratio assuming that the argon in the atmosphere is a product of the ^{40}K ^{40}Ar decay (half-life 4.5 billion years) and that ^{40}Ar is purged from the magma as it forms basalt rocks of the Earth's crust.

10. The background microwave radiation emitted by outer space has the thermal spectrum of a surface near absolute zero, 3.7 K. This is also the temperature expected of a gas that has expanded continuously over the 13.7 billion year history of the universe. In thermodynamic terms the Hubble model pictures the universe as an irreversible adiabatic expansion into a vacuum that decreases the temperature. Outer space is a virtually infinite energy sink that allows hot bodies to cool by thermal radiation.

11. Peter J. Brancazio, and A.G.W. Cameron, Eds., *The Origin and Evolution of Atmospheres and Oceans*, Wiley, New York, 1964 Selective loss of light elements from the upper atmosphere

12. Measurements have been made of the composition of the gases emitted by the Kilauea and Mauna Loa volcanoes of Hawaii. William W. Rubey, in *Geologic History of Sea Water*, Chapter 1 of Ref. 15

13. Most nitrogen is atmospheric N_2. Its cumulative mole fraction is 60.1 times its mole fraction in volcanic gases. The cumulative mass of elements outside the mantle is therefore 60.1 times their mole fraction in the volcanic gas times the present mass of the atmosphere times the molecular weight ratio.

14. If the total mass of carbonate rocks is limited by the abundance of Ca and Mg in the lithosphere, the mass of carbon in carbonate rocks is 3.42×10^{18} kg. See Peter J. Brancazio, and A.G.W. Cameron, in reference 11

15. L.V. Berkner and L.C. Marshall, *History of the Growth of Oxygen in the Earth's Atmosphere*, Chapt. 2 in *Origin and Evolution of Atmospheres and Oceans*, Peter J. Brancazio and A.G.W. Cameron, Eds., Wiley, New York, 1964

16. Gerald F. Joyce, *RNA Evolution and the Origin of Life*, Nature, **338**, 217-223, 1989

17. Stephen Jay Gould, ed., *Book of Life*, W.W. Norton, New York, 2001; Melvin Calvin, Chemical Evolution, Oxford University Press, 1969

18. William E. Galloway and David K. Hobday, *Terrigenous Clastic Depositional Systems*, Springer Verlag, Berlin, Heidelberg, 1996, Sedimentary geology emphasizing coal, petroleum, natural gas, and uranium deposits

19. The coal industry lists the production from 5000 coal mines in North America. *Coal mine directory, United States and Canada, 1989/90*, Keystone coal industry manual, McLean-Hunter Publishing Co., Chicago, 1993

20. John Scott Roy, *Birdog*, Chapman-Hall, New York, 1995 Philosophy and practice of seismic data collection and interpretation

21. M.A. Adelman, *The Economics of Petroleum Supply*, M.I.T. Press, 1993
    For example, fear that a cartel, such as OPEC, could orchestrate the price of oil withholding production from the market proved to be unrealistic.

22. G.K. Zipf, *Human Behavior and the Principle of Least Effort*, Addison Wesley, 1949; Specific examples are cited in Chapter 6 of Kenneth S. Deffeyes, *Hubbert's Peak*, Princeton University Press, 2003

23. Maynard King Hubbert, *Energy Resources*, National Academy of Sciences, Publication 1000-D, 1962; *Degree of Advancement of Petroleum Exploration in the United States*, American Association of Petroleum Geologists Bulletin, **51**;2207-27, 1967

24. 20 year old predictions about oil discovery by M. King Hubbert's theory of the growth and decline of oil production appear to remain on track. Kenneth S. Deffeyes, *Hubbert's Peak*, Princeton University Press, 2003

25. At near freezing temperature methane appears to cement water into an ice-like compound particularly at ocean pressures exceeding 300 atmospheres. U.S. Geological Survey, Woods Hole, MA, attn: Dr. William Dillon

26. Peter J. McCabe, *Depositional Environments of Coal and Coal Bearing Strata*, in *Sedimentology of Coal and Coal Bearing Sequences*, R.A. Rahmani and R.M. Flores, Eds., Blackwell Scientific Publications, Oxford, London, Edinburgh, Boston, Palo Alto, and Melbourne

27. Small coal mines use *room and pillar support* to remove coal from a room. Longwall technology with underground coal seams a mile long uses machines that load coal on a conveyor as it is dug. The conveyor moves forward as the coal face moves and the roof is allowed to collapse behind.

28. Energy Information Agency, U.S. Department of Energy, *Coal industry annual 2000*, January 2002

Chapter 4
Nuclear Power

Nuclear power began a half century ago as a practical application of the science behind the atomic bomb. It has a naturally large scale that fits the requirements of a one gigawatt base power plant. The present system is dominated by technology with a history of safe, reliable production of low cost power that is excessively wasteful of the natural uranium fuel resource. The usual scale-up method of development is not applicable. Development requires a succession of multi-billion dollar full scale plants. It has not yet converged on an industry standard. Nuclear power must ultimately meet four requirements[1].

- Provide the required base power safely and at low cost
- Optimize utilization of the natural uranium resource
- Optimize hazardous waste disposal
- Minimize risk of diversion to nuclear weapons

First generation plant designs are dominated by the first of these requirements. An industry standard would give more comprehensive consideration to all requirements from the start of planning.

Nuclear Fission

Nuclear power is based on the ability of the ^{235}U *isotope*[2] of uranium to split by *nuclear fission* when bombarded by neutrons. It forms a pair of lighter atoms plus an average of 2.5 neutrons. The fission product pairs include more than 120 different isotopes most of which are radioactive. The number of nuclear reactions is large. The discussion to follow is limited to three that are most important. Two *fissile* isotopes, ^{235}U and ^{239}Pu, react with neutrons to undergo fission. One *fertile* isotope, ^{238}U, reacts with neutrons to produce a fissile isotope, ^{239}Pu.

The key to nuclear power is neutrons formed by the fission reaction. They react with fissile isotopes to produce energy. They react with fertile isotopes to produce more fissile material.

A branched chain explosion can result if more than one fission neutron causes an additional fission reaction. This is the principle of atomic weapons. A controlled chain reaction results if just one fission neutron produces an additional fission reaction. This is the principle of atomic power.

Table 4.1 lists the specific heat components of a fission event. It releases ten million times the energy of a combustion reaction. One kilogram of uranium consumed by fission is equivalent to a 50-car trainload of coal. The 20 GWh per kilogram of heat from the nuclear reaction can supply a major power plant for 6-8 hours.

Table 4.1 Specific heat of the fission reaction

Fission products	165 MeV/ atom
Neutron kinetic energy	6 MeV/ atom
Fission gamma-radiation	7 MeV/ atom
Fission product gamma radiation	6 MeV/ atom
Fission product beta radiation	6 MeV/ atom
Anti-neutrinos	9 MeV/ atom
Energy per atom	199 MeV
Specific heat of reaction	20 GWh/kg

A heat transfer fluid must absorb the heat of the reaction and transport it to a power generator for nuclear fission to produce electric power. The fission products have most of the energy, are heavy, and transfer their energy relatively easily. They ultimately reach a concentration that poisons the reaction in a power reactor. Their radioactivity continues to generate heat after the nuclear fission has stopped. The neutrons carry the reaction further.

A critical mass is a quantity of fissile material that captures at least the one neutron per fission needed to sustain a controlled chain reaction. Fission produces an average of about 2.5 neutrons. Each has 1-2 MeV of energy. The concentration of fissile material and the physical size and geometry of the system must be large enough to capture neutrons before they escape. The fuel in electric power reactors is far below the concentration where explosion is possible. A nuclear explosion requires a compact mass of essentially pure fissile material.

A moderator is a medium surrounding the reactor that slows neutrons by elastic collisions. *Thermal neutrons* have energies of less than one eV. Heavier atoms require more collisions to reach thermal energies. The hydrogen atoms in water are an effective moderator because they have the same mass as neutrons. However hydrogen atoms ^{1}H also capture neutrons to form deuterium ^{2}H.

Deuterium oxide or *heavy water* is an even more effective moderator. The probability of neutron capture to form tritium ^{3}H is small. Most collisions simply divide the energy.

Neutron capture is possible for some isotopes of all elements. Fissile elements undergo fission by capturing neutrons at all energies. Although fissile isotopes can capture high energy neutrons the probability is greater with lower energy neutrons.

Neutron capture by non-fissile elements produces a heavier isotope that may or may not undergo radioactive decay. Neutron capture by fertile ^{238}U to form fissile ^{239}Pu is the key example for nuclear technology. The capture probability strongly favors lower energy neutrons.

The fuel for an atomic weapon typically reaches a critical mass at 3-5 kg of pure plutonium metal or 15-20 kg of 95% ^{235}U metal. The critical mass also requires a critical geometry. A fast branched chain reaction requires a small mass, small volume, and nearly pure fissile material. Since the kinetic energy of the reaction products immediately disperses the system. The critical mass must be assembled with explosive speed. Favorable reaction conditions exist only momentarily. The efficiency of the reaction is low.

Nuclear Facilities

Fig. 4.1 shows the typical facilities of a nuclear program[3]. Weapons technology and national security dominated the original technology and continues to leave a heavy imprint even though it is a small fraction of the total production.

First generation nuclear power plants use the technology of the heavy boxes down the left side of the figure. The facilities in the center and right hand columns produce fuel for plutonium and uranium weapons technology, respectively. Second generation nuclear power plants that are more fuel efficient are likely to use a derivative of the technology used to produce plutonium.

The fuel in present generation electric power reactors is natural uranium with the ^{235}U enriched from its natural 0.7% to about 3.5% ^{235}U. To enrich the uranium it is first produced as a gas in the form of the high vapor pressure of UF_6 crystals[4].

Gas centrifuges subject the UF_6 gas to a gravitational force of nearly a million times normal gravity to separate the ^{235}U. About 20 units in series produce 3.5% separation. Many units operating in parallel are necessary for significant production. Enrichment facilities use thousands of individual units[5]. The total energy from a reactor charge increases in proportion to the enrichment up to 20-30%. The enrichment cost increases geometrically[6].

Pressurized water reactors account for over half the existing power plants[7]. The reactor core is typically an assembly of several hundred fuel rods containing 25,000 kg of uranium enriched to 3.5% ^{235}U. The *fuel rods* contain UO_2 ceramic pellets in metal cladding. The critical mass is sufficient to produce 1 GW of electrical power while consuming most of the fissile material over an 18-24 month period. A chain reaction starts when the array is immersed in light water moderator. The reactor is a high pressure vessel containing water heated to a super-critical temperature and pressure. Control rods of neutron absorbers control the power level. The core continuously transfers heat to un-radiated water working fluid to drive a steam turbine engine.

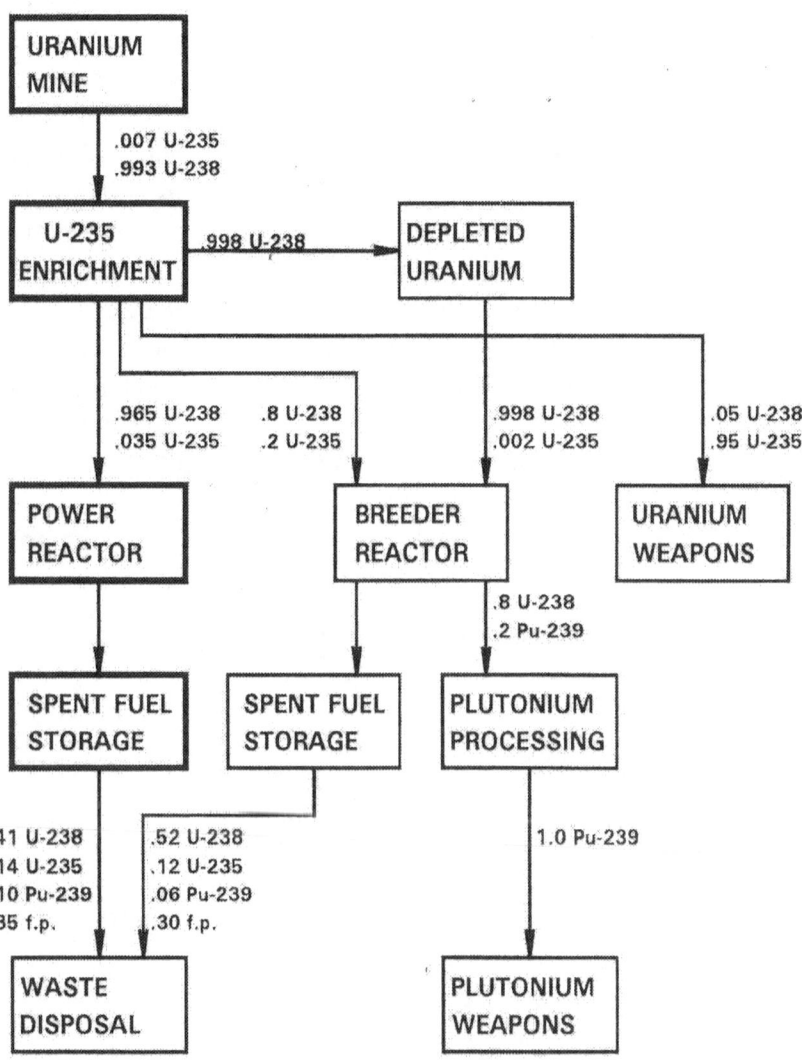

Fig. 4.1 Disposition of nuclear materials in nuclear power and weapon facilities

A good reactor design has temperature coefficients that spoil the critical mass and stop the reaction when there is excessive heating due to an accidental loss of coolant. Late in the fuel cycle the heat from radioactive fission products is sufficient to cause a *melt-down* even without a nuclear reaction

Concrete containment 2 meters thick isolates the reactor and radioactive components and captures neutrons and γ-radiation not stopped by the water. A fuel charge provides 1 GW nameplate power on an 18-24 month fuel replacement cycle.

Table 4.2 shows the material balance in a pressurized water reactor fuel cycle[8]. The fission products from one fuel cycle correspond roughly to the initial charge of fissile material. The reactor charge absorbs 1.28 neutrons per initial fissile atom, the ratio of all neutron reaction products to fission products. The 0.28 neutrons produce enough fissile ^{239}Pu to maintain a constant reaction rate as the original fissile material is depleted. Some fissile material remains in the spent fuel.

Table 4.2 Reactant-product balance in a power reactor cycle

Fission products	946 kg
Pu-239/240 residue	266 kg
U-235 residue	391 kg
U-238 residue	25,655 kg
Initial enriched uranium	27,282 kg
Depleted uranium	152,800 kg
Natural uranium input	180,000 kg

Uranium Supply and Demand

The source rock for uranium deposits is granite formed by hydrological flows in basalt containing 3 ppm uranium. The flows that change basalt to granite also transport minerals out of the granite through other rock formations that selectively adsorb and concentrated them in separate deposits. Uranium deposits are found in at least a dozen different kinds of geological formations, usually as uranium oxide U_3O_8.

Most known deposits are small enough to be completely mined out in a relatively short time. The size of known deposits decreases rapidly from a maximum of about 50 kt of uranium. A few deposits have concentrations as high as 20%-50% U_3O_8[9]. The ore from more typical mines is in the 0.2 to 1% range at a cost related to the quantity and kind of rock which must be processed.

To recover the pure uranium the ore is crushed and ground to a powder which is then leached with a fine spray of dilute sulfuric acid. This concentrates the ore to 50% U_3O_8 in a form called *yellow cake*. To purify the U_3O_8 it is converted to a chelate of acetone in water solution. The chelate is extracted in a hydrophobic solvent, crystallized, and heated in an oven.

The uranium reserves associated with deposits that have been discovered and developed are classified according to the cost of mining and production as yellowcake. The cumulative total quantities are 943 kt at less than \$40/kg, 3085 kt at less than \$80/kg, and 4299 kt at less than \$130/kg. Uranium reserves must be viewed as the exploration, development, and production associated with a limited history of the demand of an emerging electric power market.

Nuclear power must increase very substantially for it to play a major role in the world electric power demand. Nuclear power, coal, and hydroelectric power are the only presently available sources with the capacity and reliability to provide the base power for the grid. In nations without other energy sources, such as France, Japan, and South Korea, it already supplies a major share of the base power. A factor-of-3 increase in the present 19% share would provide more than half the present world demand.

The present demand from the 438 operating nuclear power plants totals nearly 400 GW. For the purpose of discussion the baseline future demand for all electric power energy is 5,000 GW_e or 15,000 GW_h. The scope of electric power needed to meet the demand is under-appreciated. The annual baseline demand would consume more than double the fissile materials in the world's nuclear weapons stockpiles[7]. This is the energy equivalent of 75 atomic bombs per hour!

71

Present technology is near the end of its first generation power plants. Pressurized water reactors are straightforward and safe but they use only the 0.5% of the natural uranium resource that can provide enrichment in ^{235}U. Much of the ^{238}U remainder is unusable. This was tolerable while uranium was cheap and the demand was modest. Better use of the available resource requires new generation of reactors.

Fast Breeder Reactors[11]

To avoid wasting the natural uranium resource a reactor must have two independent functions. One is to generate electric power. The other is to produce fissile plutonium reactor fuel.

The power reactor function takes place in a core similar to a pressurized water reactor. The fuel rods contain uranium enriched with fissile material. They sustain controlled fission to produce power. They also provide extra neutrons to produce plutonium. To do this the fission reaction uses fast neutrons with little moderation. Liquid sodium metal is a poor moderator and poor neutron absorber and does not significantly interfere with the neutrons. Like other liquid metals the high thermal conductivity of sodium has excellent heat exchange properties. Most of the neutrons that are not consumed to sustain the reaction escape from the core.

The breeder function takes place in breeder fuel rods that surround the core. They contain fertile ^{238}U. This can be either uranium depleted by the enrichment process or natural uranium. It is in ceramic pellets of UO_2 encased in rods of metal cladding like the reactor fuel but larger in diameter. This produces a ^{239}Pu mixture that is not greatly contaminated by fission products. Chemical extraction produces pure ^{239}Pu with only modest concern for radioactivity.

^{235}U fission produces 2.5 neutrons and ^{239}Pu fission produces even more. The geometry and composition of the reactor must maximize the number of neutrons per fission that produce ^{239}Pu. If the breeder reaction produces as many ^{239}Pu atoms as the fissile atoms the power reactor consumes, ^{235}U enrichment is unnecessary.

Breeder fuel rod contents provide the fissile material for the power reactor fuel rods. The reaction products of the breeder rods are a mixture not greatly different from the fuel required for the power reactor. It should be possible to reduce processing to a minimum.

Breeder reactor fuel rods are a mixture of 20-30% ^{239}Pu in ^{238}U known as *mixed oxide MOX fuel*. The total energy from a given core corresponds to complete consumption of the fissile material. It increases with the ^{239}Pu content up to a point. There is a compromise between maximum consumption and the *breeding fraction*, the fraction of neutrons that produce ^{239}Pu. As the reaction progresses the power reactor fuel rod contamination poisons the efficiency of both the power reaction and the excess neutron production.

Fast breeder reactors consolidate many of the functions described in Fig. 4.1 when they are used for electric power. In principle the plutonium processing output can eventually replace a need for enriched uranium input. Uranium enrichment facilities become unnecessary. The breeder reactor and plutonium reprocessing form a closed loop except for the spent fuel storage and ultimate radioactive waste disposal. Eliminating the separate enrichment facility with its huge array process streams containing fissile materials consolidates the distribution of fissile materials that might be misappropriated for weapons.

A successful breeder reactor meets the primary objective by increasing the utilization of the natural uranium resource by up to a factor of about 40. The fissile atom content of natural uranium sets the utilization in pressurized water reactors at an upper limit of 0.5%. The maximum fissile fuel content that can still achieve burn-up sets the utilization by breeder reactors at an upper limit of 20-30%.

Radioactive Waste

All materials associated with a nuclear reactor ultimately become either spent fuel or a lower radioactive waste hazard. Radiation exposure is measured by some combination of radiant power, exposure time, and the penetration properties of the specific type of radiation.

It is usual to describe the high energy of radioactive radiation as radiation particles whether they have mass or are simply high energy photons of electromagnetic radiation. Alpha, beta, gamma, neutron, and cosmic radiation particles have different penetration and capture properties according their individual energy, mass, and charge.

The linear-no-threshold principle implies that radiation is dangerous at all levels. This is both unrealistic and untrue. It may prevent abuses of radioactive hazards but it also promotes panic in the event of accidental exposure. The radiation hazard depends on the particular circumstance. People who work routinely in a radiological environment must understand the hazard.

Radioactivity decreases over time but never disappears. Natural uranium is the remains of 4.5 billion years of decay. It is one of several exposures that are ubiquitous, unavoidable, and not necessarily harmful. Among survivors of Hiroshima and Chernobyl a range of low exposures was not harmful and possibly even beneficial[13]. Biological mechanisms repair damage that is not too critical or massive. In the U.S. almost 17 million medical procedures per year inject radioisotopes. The hazard from substances like ^{90}Sr, ^{137}Cs, and most isotopes of plutonium is both chemical and radiological.

Spent fuel, the most concentrated radioactivity since the planet was formed, is too complex to characterize. Nuclear reactions make it a thermal hazard that must cool for a year to handle. It is mostly ^{238}U with a small fraction of fission products, and still smaller fraction of fissile isotopes. The fissile isotopes are contaminated with isotopes ^{232}U through ^{239}U and ^{232}Pu through ^{246}Pu. These are radioactive and not easily separated. Spent fuel with 20% fission products has no value worth recovering by processing.

Table 4.3 calculates the spent fuel storage required for a possible North American power demand of 600 GW_e. The spent fuel output of fast breeder reactors would fill the space planned for the Yucca Flat Nevada repository in 72 years. This is over 100 times smaller than the volume of the ore used to produce the uranium. Lower order waste would require proportionately greater space.

Table 4.3 Radioactive waste for 600 GW$_e$

Heat of fission	20 GWh/kg
Reactor fission products	438 kg/GWy
North America nuclear heat	1800 GW
Total fission products	788 t/y
Total spent fuel, pressurized water	22525 t/y
Total spent fuel, breeder	3942 t/y

Structural and process equipment, such as the radioactive remains of the plant after its useful life, has a range of lower order radiation hazards. They are different enough to reprocess selectively. Stainless steel too radioactive for public consumption is suitable to reprocess as the waste containers that are buried. These would be protected by temporary shipping containers.

The liquid sodium heat transfer fluid eventually becomes radioactive. Ideally it would be separated into high level radioactive waste and reprocessed heat transfer fluid.

Waste process solutions are simple to purify and recycle. An arbitrarily small concentration of radioactive material always remains. At some point thousands of gallons become too dilute to recycle. These can only be discarded as ordinary waste at a partly arbitrary fraction or multiple of the background radiation.

The important point is that every class of radioactive waste requires a policy decision to address and not leave uncertain.

Nuclear Weapons Security

Plutonium is relatively easy to purify and convert to weapons grade material by ordinary chemical processes. A bomb requires a relatively small amount of pure metal. Greater plutonium production intensifies the need for accountability and control of material that can produce a nuclear weapon.

On the other hand, fast breeder reactors concentrate the danger points that must be protected. The greatest exposure of bomb grade material occurs at the point of converting the products of the breeder reactor to fuel rods for the power reactor. Both the breeder tubes and the fuel rods must be sealed, counted, and guarded.

Simplicity is worth a premium. Fast breeder reactors offer an opportunity to consolidate and simplify the operations and minimize the processing required to produce nuclear reactor fuel in a combined power generation-fuel production facility. The material in the irradiated breeder tubes is similar to that in the fuel rods except for the plutonium concentration.

Spent fuel is a different kind of weapons risk. Given the large quantity that would have to be processed and the extreme difficulty and hazard in processing it diversion to produce a bomb is unlikely. A more likely use would be as shrapnel to spread the radiation by ordinary high explosives in a *dirty bomb*. In this case panic from the inordinate fear of radiation that has been instilled in the population could create as many casualties as the radiation exposure and explosion.

Ideally nuclear power facilities would be built at sites with stable geology and deep waste burial that is impervious to water. This would minimize transporting nuclear materials and consolidate security.

Notes and references for chapter 4

1. P.D. Wilson summarizes some specific suggestions for improving past practices in *Chapter 14 Future Perspectives*, in *Nuclear Fuel Cycle from Ore to Waste*, P.D. Wilson, ed. Oxford University Press, 1996
2. The superscript preceding the symbol for an element is the sum of the number of protons and neutrons. It is the approximate atomic weight of a particular isotope of that element.
3. A description of the world's nuclear power facilities is given in *The World Nuclear Handbook*, Euro monitor Publications, Ltd., 1988.
4. Anhydrous HF gas at elevated temperature reacts with U_3O_8 to produce UF_4 (green salt). Further reaction with F_2 gas yields UF_6, a white crystalline solid with a low vapor pressure.
5. Gas diffusion first produced the initial enriched uranium in a massive plant at Oak Ridge Tennessee. Other plants at Tricastin France and four sites in Russia can each supply 100 power reactors. They enhance uranium to 3.5% in 1000-1400 stages. Gas centrifuges are 50 times more efficient.
6. The cost of enriching uranium is measured in *separative work units* or SWU. About 5 SWU enriches 1 kg of natural uranium to 3.7% ^{235}U while depleting 6 kg to 0.2%. The cost depends on the electrical energy consumed, about 100 kilowatt-hours per SWU. This adds roughly \$5/kg to the cost of natural uranium for a reactor using 3.7% ^{235}U.
7. John Lillington, *Future of Nuclear Power*, Elsevier Press, 2004 describes the design of many of the types of reactor that have been used.
8. H. Nifenecker, S. Orvid, J.N. Loiseaux, and A. Giorni, *Hybrid Nuclear Reactors*, Institut des Sciences Nucleaire, Grenoble, France, 1998
9. Uranium deposits in Gabon Africa have fission products that are evidence of a critical ^{235}U concentration about 2.2 billion years ago that maintained chain reactions periodically for many million years. At that time the ^{235}U natural abundance was about 5 times its present value. Deposits are more typically 0.2% uranium.
10. The world stockpile of weapons grade nuclear materials is estimated to be 470 t ^{239}Pu and 2300 t ^{235}U. This corresponds to about 440 kt of natural uranium.
11. A.M. Judd, Fast Breeder Reactors, Pergamon Press, 1981
12. The baseline requirement using pressurized water plants with no reprocessing corresponds to over 100,000 kt for each reactor fuel cycle. International Atomic Energy Agency estimates reasonably assured uranium reserves of 4400 kt.
13. Edward J. Calabrese and Linda A. Baldwin, *Environmental Health Perspectives*, **106**, 357-362, 1998, cite 4000 publications.

Chapter 5
Coal, Oil, Gas, and Nuclear Economics

An efficient electric power market must resolve multiple complications. All electric power has the same value. The only quality issue is whether the supply is reliable. The efficient market price of electric power would depend only on the real cost to produce and distribute it. This depends on the technology.

Power distribution companies have a natural monopoly that comes from owning and maintaining the physical connections to consumers.

Base power generators have a natural monopoly. The low cost requires a scale that dominates a region. Inter-regional power grids and transmission lines increase the cost and decrease the reliability. Present technologies face the likely prospect of long term declining supply and increasing cost. New technology requires operating experience over many generations of systems and design improvements. New technology is unlikely to satisfy the required demand without capital costs that require 20-30 years to amortize.

An economically feasible power plant must generate revenue that covers the costs. *Revenue* is the product of three factors. The *nameplate power* is the maximum power the plant can produce at peak demand plus a margin of safety. The *utilization factor* is the average fraction of the nameplate power the plant actually delivers over time. The *price per unit of energy* must give revenue that balances the total cost. *The capital cost* includes returning the original investment with interest.

Coal and nuclear power generate the present base power at a competitive lowest price. Supplying both the peak demand and the minimum demand with high utilization is a contradiction. The lowest cost trade-off is now a combination. Base power facilities generate most of the power at low prices. Peaking power facilities generate power during short periods of maximum demand with lower utilization and higher prices. Natural gas is the fuel for most peaking power facilities.

Large power plants require planning to meet the demand with high reliability. Each uncertainty in planning power delivery raises the cost of power. To produce power at the lowest cost the total capacity should match the total maximum demand plus a margin of safety. Reasonable expectations of a market mechanism are not easy to define.

This chapter presents tables that calculate a plausible balance of revenues and costs for the major current electric power technologies. Experience provides a basis for the capital costs. Later chapters calculate similar tables for targets emerging technologies must meet. These calculations take care to avoid prejudging the ingenuity of designers.

Economics of Power from Coal

Table 5.1 shows the arithmetic of economics that reconciles revenues from coal technology with the capital and operating cost. For coal the annual payments for fuel, capital, and other operating costs are each roughly a third of the total.

The revenues are $315 million for a 1-GW plant operating at 60% utilization at an average price of 6 cents per kilowatt hour.

The capital cost of $1500 per kWh of nameplate power is a value typical of coal fired power plants[1]. The return on investment assumes that the capital is repaid at a uniform rate over the amortization period with interest on the remaining investment. The total repayment is roughly double the initial investment.

The capital cost includes equipment to move the coal bed through the furnace, a high-pressure steam boiler, and one or more steam turbine engines driving electric power generators. It excludes most of the cost of environmental remediation that is increasingly necessary as the quality of coal resources declines.

The operating cost is divided about equally between the fuel cost and other operating costs. The other costs include maintaining the machinery to transport and burn 50 to 100 carloads of coal each day.

Coal is chemically similar to pure graphite carbon. Its heat of combustion, $\Delta H=-393.5$ kJ/mole, is derived from the reaction,

$$C(graphite) + O_2(air) = CO_2(gas) + H_2O(gas)$$

The 7287 kWh/t is the heat content of Western bituminous surface mined coal that is 88% pure carbon. The fuel cost includes an allowance for transportation from the mine and disposal of fly ash removed in the environmental treatment.

Table 5.1 Economics of electric power from coal

Revenues	
Nameplate power	1 GW
Utilization	0.60
Average price	$0.06 /kWh$_e$
Electric energy	5256 GWh/y
Annual revenue	$315 million/y
Capital Cost	
Amortization period	25 y
Return on investment	0.08
Payment on capital	$120 million/y
Plant replacement cost	$1500 million
Operating Cost	
Heat-to-Electric efficiency	0.32
Annual thermal energy	16425 GWh
Heat value	7287 kWh/t
Coal quantities	2.25 million t/y
Coal price delivered	$40 /t
Total fuel cost	$90 million/y
Other operating cost	$105 million/y

A decision to build a new plant requires planning that guarantees revenue through the amortization period. Some mines might guarantee 2 million metric tons of coal per year to a single 1 GW plant for a decade, or so. Few, if any, can guarantee delivery at that rate and quality for 25 years. The alternatives are some combination of smaller deposits, lower quality, and more difficult access.

The largest deposits of high quality coal that can be mined from the surface will disappear first. The operating coal face of some mines with very large reserves hundreds of feet below the surface is already several miles away from the entry shaft.

Air pollution by power plant emissions is an increasing factor in the cost and future use of coal. The 0.5% organic sulfur and 1.5% organic nitrogen in high sulfur coal contributes to the aerosol smog that is visible in most highly populated regions. The smog contains sulfuric acid and nitrogen oxides plus part of the 5-10% fly ash. The ash contains the natural abundance of most metals, including heavy metals such as mercury[2]. Some are more radioactive than nuclear power plant emissions. The sulfur contributes 20,000 t/y of atmospheric sulfuric acid rain from each power plant.

Most of the carbon in coal becomes carbon dioxide emission. The concern requires the more complete discussion in Chapter 6.

Present environmental remediation technology removes pollutants by treating the flue gas. The volume of gas combustion products is roughly a million times the volume of the coal. The volume ratio alone dictates a process that is inherently inefficient and expensive.

Emerging remediation technology emphasizes methods of cleaning the coal before it is burned. The first level of treatment is to pulverize the coal to a quasi-fluid. Injecting it mixed with air into a burner can produce a combustion temperature high enough for a high temperature gas turbine. This improves the heat efficiency in comparison with lower temperature steam turbines.

Combined cycle technology might be an advantageous way to use the gas turbine exhaust to heat steam. Combining a gas turbine with a steam turbine might raise the thermal efficiency to as high as 50%.

A combustion chamber design might remove most of the fly ash by centrifugal separation. This does not resolve pollution from sulfur and nitrogen oxides.

Integrated gasification is a technology with better potential for pollution abatement. *Synfuel* is a generic term for producing combustion fuels by refining other feed stocks. Coal refining produces *syngas*, a mixture of hydrogen and carbon monoxide.

Table 5.2 compares the synfuel reactions with the coal combustion reaction. Note that the sum of the combined synfuel reactions is identical to the direct combustion of coal, 18.22 kWh/kg.

Table 5.2 Comparison of Syngas and Coal combustion

	$kWh_h/kg\ C$
Synfuel reactions standard heats of reaction	
$C(graphite) + \frac{1}{2}\,O_2(g) = CO(g)$	-2.56
$C(graphite) + H_2O(g) = H_2(g) + CO(g)$	3.04
$2\,CO(g) + O_2(g) = 2\,CO_2(g)$	-13.10
$H_2(g) + \frac{1}{2}\,O_2(g) = H_2O(g)$	-5.60
Total	-18.22
Coal standard heat of combustion	
$2\,C(graphite) + 2\,O_2(g) = 2\,CO_2(g)$	-18.22

Syngas is the reaction product by injecting pulverized coal feedstock into the bottom of a reactor and a carefully controlled mixture of oxygen and steam in the top. The key reaction of steam with graphite to form syngas cannot occur spontaneously at ordinary temperature. The large entropy increase caused by the increased fraction of gas molecules gives a favorable free energy at high temperature.

The countercurrent flow of oxygen and steam from the top and carbon from the bottom creates a stratified reaction zone[3]. Heat is extracted in the mid-portion of the reactor at the optimum temperature for the working fluid of a gas turbine engine. The highest temperature is at the bottom where the coal first meets the oxygen.

Absorbents that are added to remove sulfur act as flux that melts the fly ash. Liquid slag that contains most of the pollutants can be drained from the bottom.

Economics of Power from Petroleum

Petroleum has base power reliability. Its value as fuel for transportation is too high for it to produce competitive electrical base power. It is used mostly in smaller, older power plants in the U.S. Northeast coast where factors such as lower pollution, high population density, high land cost, and access to ocean transport and shipboard storage compensate for the high cost.

Table 5.3 shows the arithmetic that balances revenues, capital cost, and operating cost. The revenues for this example are based on a smaller 0.1 GW plant more typical of existing plants than the 1 GW scale example used for coal. New oil-fired plants are not being built.

The capital cost of an oil-fired plant is typically $1000 per kilowatt-hour. The simpler oil burners, steam boilers, and fuel handling equipment makes the cost about 2/3 that of a coal fired plant. This also allows lower non-fuel operating cost. Environmental remediation by refining the crude oil is more effective than removing pollutants from the flue gas.

The cost of fuel dominates the economics of petroleum. Fuel, capital cost, and operating cost contribute about equally to the cost of electricity from coal. As the price of oil rises through the $40 to $80/bbl range fuel begins to be a more dominant share of the price of power. A petroleum price of $80/bbl requires an electricity price of 21 cents/kWh to give enough revenue to pay for the capital cost and minimal operating cost.

At $80/bbl the cost of petroleum is about 90% of the price of the electric power it produces. Petroleum illustrates one aspect of the end of fossil fuel resources. The cost of fuel changes from one of several comparable costs of production to become the dominant cost.

This is a fundamental transition in the view of the value of natural resources that is necessary. It is the point at which continuing use of power no longer depends on simply discovering new deposits of old resources.

Table 5.3 Economics of 0.1 GW of electric power from petroleum

Revenues

Nameplate power	0.1 GW
Utilization	0.60
Price	$0.21 / kWh$_e$
Electric energy	525 GWh/y
Revenue	$94.6 million/y

Capital cost

Amortization period, yrs	25
Rate of return on investment	0.08
Capital cost payment	$9.5 million/y
Total plant cost	$100 million

Operating cost

Electric/heat efficiency	0.32
Annual thermal energy	1642 GWh/y
Heat of combustion	11600 KWh / t
Oil quantities	141 thousand t/y
Oil price	$80 / bbl
Oil price	$672 / t
Total fuel cost	$95.1 million/y
Other operating cost	$7.1 million/y

Fuel oil has many different compositions depending on its source and how it is refined. It consists predominantly of straight chain hydrocarbons. The energy density of petroleum is about 50% greater than coal due to the hydrogen content of hydrocarbons. The heat of combustion of dodecane, $\Delta H = -7575$ kJ/mole, is typical. It burns in accordance with the reaction,

$$C_{12}H_{26} \text{ (liq)} + 25 O_2 = 12 CO_2 \text{(gas)} + 13 H_2O \text{ (gas)}.$$

Several factors combine to make the price of petroleum too high for electric power. The competition by transportation makes coal and nuclear energy cheaper. The extreme demand by transportation makes the price of petroleum greater. The price depends on the discovery, development, and production cost of the most expensive petroleum. The rate at which less expensive petroleum can be pumped sets the demand for the higher price petroleum.

Economics of Power from Natural Gas

Natural gas power plants are designed for relatively short periods of peak demand. They use different technology than coal or oil. The gas is burned directly in the combustion chamber of large gas turbine engines that can be brought to full power from a cold start in about 15 minutes, as opposed to many hours.

Table 5.4 shows the balance between revenues and expenses for a 0.1 GW gas fired power plant. This size is consistent with the relatively smaller utilization in comparison with base power. The price is set at a value that yields enough revenue to cover the fuel, operating, and capital costs.

Table 5.4 Cost of electricity from natural gas

Revenues	
Nameplate power	0.1 GW
Utilization	0.15
Price	$0.18 /kWh$_e$
Annual electric energy	131 GWh
Annual revenue	$23 millions
Capital cost	
Amortization period	20 y
Rate of return on investment	0.08
Capital cost payment	$6.7 million
Total plant cost	$75 million
Operating cost	
Heat-to-electric efficiency	0.35
Annual thermal energy	375 kWh
Heat of combustion	13900 kWh/t
Gas quantities	27 thousand t/y
Gas price	$504 / ton
Total fuel cost	$13 million / t
Total operating cost	$3.2 million /y

Gas turbines can be considered emergency generators that are activated when the spinning reserves of the base power plants become low. Their cost is a premium added to the value of power in the peak demand period. Charging this premium directly to individual consumers by time-of-use metering is now standard practice.

The capital cost of a natural gas plant is about $750 per kilowatt-hour of nameplate capacity. The lower capital cost results mostly from the greater simplicity of the fluid flow at the gas turbine generators. Combustion avoids the high pressure boilers system. The higher combustion gas temperature adds to the thermal efficiency of the engine.

Aside from the engine technology the major difference with respect to petroleum is the low utilization. Under these conditions each percentage of increased utilization adds a little under a million dollars to the non-fuel operating budget.

The composition of natural gas is a combination of methane, ethane, and propane depending on the source. The energy basis used in Table 5.4 is methane. Its heat of combustion is $\Delta H = -802.7$ kJ/mole corresponding to the reaction,

$$CH_4(gas) + O_2(air) = CO_2(gas) + 2\ H_2O(gas)$$

Natural gas is the fuel of choice for home heating as well as peaking power. Both uses fill an essential need for which there are no good substitutes.

Gas supplies are more ubiquitous than petroleum. The high growth in recent gas utilization and pipeline delivery has not yet revealed the vulnerability of supplies that have become characteristic of petroleum.

Economics of Nuclear Power

Table 5.5 summarizes the economics of nuclear power on the same 6-cent per kWh revenue basis used for coal. Nuclear energy is a low cost and high reliable fuel that is competitive with coal source of base power.

Nuclear power reactors have even greater inertia and smaller range of spinning reserve than coal fired power. The high utilization assumes that a nuclear power plant will be able to sell large quantities of off-peak power for transportation replacement applications such as electrolytic hydrogen generation.

Table 5.5 Economics of electric power from uranium

Revenues	
Nameplate power	1.00 GW
Utilization	0.70
Price	0.06 $/kWh
Annual electric energy	6132 GWh / y
Annual revenue	368 $million / y
Operating cost	
Electric/heat efficiency	0.32
Annual heat energy	19162 GWh
Heat of combustion	20 GWh/kg
Fission products	958 kg / y
Natural uranium	191625 kg / y
Uranium price	$80 /kg
Total fuel cost	$15.3 million / y
Total operating cost	$192.7 million / y
Allowable capital replacement cost	
Return on investment	0.08
Amortization period	25 y
Capital cost repayments	$160 million
Plant replacement	$2000 million

The finances are calculated for a unit with 1.0 GW nameplate power operating at 70% utilization for 25 years. The capital cost is based on the $2000 per kW typical of recent pressurized water plants. A *pressurized water reactor* captures the heat of a nuclear reaction to generate electric power by a conventional steam turbine. The high utilization assumes time-of-use price contracts.

The life of a pressurized water plant is limited by the solubility of stainless steel at the high temperature and pressure of the supercritical steam. Aggravated by radiation it erodes inches from the walls of a stainless steel reactor over its lifetime. Radiation limits the possibilities for corrective maintenance and replacements that are standard practice in usual steam plants. The reactor must be extremely robust for safety over its lifetime.

Fast breeder reactors require new development that is ultimately necessary for a low cost, maximum lifetime power industry standard. It may not be the immediate lowest cost nuclear power to install. The economic questions would quickly disappear if a clearly superior industry standard fast breeder reactor were to appear. More realistically, these developments more often occur incrementally in an environment where new reactors are continually under construction.

Waste disposal and security costs are not fully covered by the revenue cited in this account. Ultimately deep underground waste storage is a necessary cost of commercial power. Waste disposal facilities located near the waste generating facilities encounter fewer political and security problems. This may be a consideration in choosing sites for power facilities.

Notes and references for chapter 5

1. The capital cost of each technology is the initial equity capital, or plant replacement cost. The required annual payments assume that the capital is repaid in equal annual increments plus the interest on debt remaining at the start of each period. Note that payments at 8% over a 25 year amortization period total double the plant replacement cost.

2. The top layers of a coal deposit contain increasing mineral content. This contains sulfur minerals and heavy metals. Although radioactive minerals in fly ash are minor, they emit more radiation than well designed nuclear power plants.

3. A syngas reactor resembles a blast furnace in size, temperature, and ceramic lining. The product is also used as a petroleum replacement feed-stock for chemical synthesis.

Chapter 6
Electric Power and the Environment

The *U.N. Framework Conference on Climate Change* met in Kyoto, Japan, in December 1997 to formulate an international treaty to limit carbon dioxide emissions from burning carbon based fuels[1]. The treaty asserts that the increase in atmospheric concentration of CO_2 from fossil fuel consumption is causing global temperature to increase at an alarming rate. The fear is that it will reach a tipping point of irreversible climate change.

Economic[2] and political[3] overtones cloud the science. The explanation of the mechanism of the change is not clear. If the danger is as serious as claimed it is not clear whether the treaty is an adequate remedy.

It is counterintuitive to consider atmospheric CO_2 a threat to the environment. It provides all of the carbon for all life on Earth. It is the food that enabled land plants to evolve and create an oxygen atmosphere. Plants are the food for animals.

Increasing the concentration of CO_2 increases the growth rate of plants, particularly trees in their prime growing age[4, 5]. The CO_2 probably had 10-20 times greater concentration than at present during the carboniferous era of peak plant growth[6]. Oxygen is the product of plant growth. The resulting plant life created both the oxygen atmosphere and the fossil fuel that is now a concern.

It is also counterintuitive to think of greenhouse gases as a threat. Planets with little or no atmosphere, like Mars or the moon, change temperature by well over 100 degrees C as they heat toward a high local solar *radiation temperature*[7] during the day and cool toward the minus 269 C temperature of outer space at night. The Earth's climate is moderate by comparison with daily temperature changes by only about 20 C over a wide temperate region.

Carbon dioxide is a transparent gas. Water vapor is another greenhouse gas. However water condenses in the atmosphere to form the clouds that are normally accepted as the visible signature of weather and climate.

Heat Trapping by the Atmosphere

The textbook example of the greenhouse effect is Venus. Its atmospheric temperature is 460 C, roughly the local solar radiation temperature. Its atmospheric pressure is 100 times that on Earth. The Venus atmosphere is almost pure CO_2, 2500 times the concentration in the Earth's atmosphere. This makes the CO_2 greenhouse effect on Venus 250,000 times greater than on Earth.

A greenhouse effect traps radiant heat in an atmosphere that is transparent to sunlight. The atmosphere absorbs infrared thermal radiation that would otherwise be radiated to space. Infrared excited states emit radiation spontaneously within a few milliseconds. To store a substantial quantity of heat the *optical thickness* of the atmosphere must be great enough to absorb and re-emit energy many, many times before it escapes.

Fig. 6.1 compares the spectral radiant power striking the upper atmosphere with that reaching the Earth through a clear sky. The hemisphere of the Earth exposed to daylight intercepts 173 PW of solar power at 1.37 kW/m^2 radiant power density. This is called the *solar constant* although is not entirely constant. The Earth scatters or reflects about 52 PW as the albedo and absorbs 121 PW.

The upper spectrum is the spectral radiant power of the sun at an altitude above the atmosphere[8]. The lower spectrum is the spectral radiant power of the sun reaching the surface through a clear sky at 30% relative humidity[9]. This is energy that heats the Earth.

The difference between the two spectra in the visible and ultraviolet regions is partly the result of absorption by electronic energy states of oxygen and ozone. It is partly the result of molecular scattering which increases in proportion to the 4th power of spectral frequency.

The difference in the infrared is due to absorption by molecular vibrations of water, except for a weak CO_2 band near 5000-cm^{-1}. Water vapor absorbs about 14% of the solar radiant power directly at this humidity.

About 70% of the Earth is covered by clouds at any given time. The complete account of this energy is obviously much more complex than the clear sky interactions.

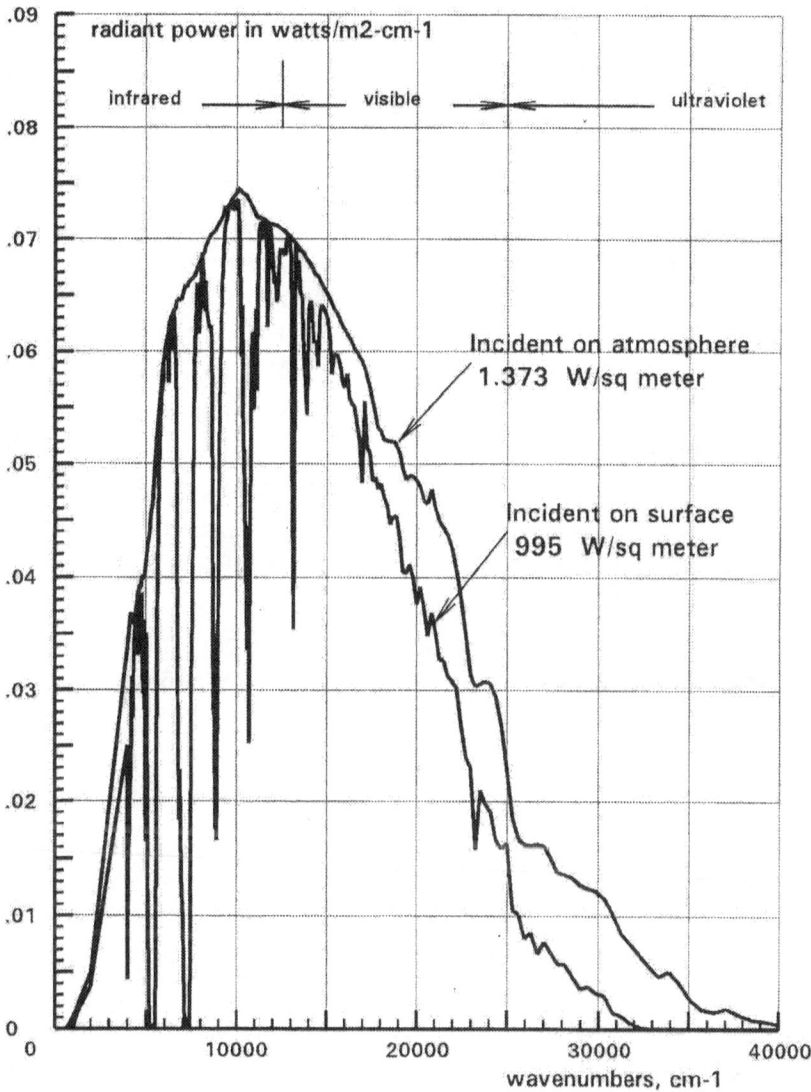

Fig. 6.1 Solar spectral radiant power above and below the clear sky atmosphere

The solar constant has a periodic increase at about 11-year intervals associated with sunspots[10]. *Sunspots* are the dark centers of much larger, bright disturbances that increase the total radiation by about 1% and eject large volumes of charged particles[11]. Earth's temperature fluctuations correlate more strongly with solar emission than with changes that can be attributed to CO_2[12]. The fluctuations are not large enough or long enough to be significant.

To maintain the energy balance the energy absorbed by the Earth must be lost by thermal radiation to space. This occurs continuously over the total area of the planet. The 121 PW input is equivalent to average thermal radiation at 254 K or -19 C.

Fig. 6.2 compares the spectrum of thermal re-radiation at typical atmospheric temperatures with absorption by CO_2 and H_2O. The atmospheric greenhouse effect is not reflection, as in a glass greenhouse, but radiation transfer[13]. A large infrared greenhouse effect requires an optically thick atmosphere which absorbs and re-emits the radiation many times before it escapes to space. To emit and absorb radiation the motions of a molecule must create an oscillating dipole moment. A multitude of low frequency vibrations give liquids and solids a continuous infrared spectrum that accounts for strong emission from cloud tops.

H_2O vapor has a permanent dipole moment that gives an exceptionally strong rotational spectrum between 0-400 cm^{-1}. The bending vibration at 1595.0 cm^{-1} is much weaker. Vibrating molecules also rotate. The spectrum at a vibration frequency appears with a band of rotational lines on either side depending on whether the transition adds or subtracts rotational energy.

CO_2 gas is a symmetrical linear molecule with no permanent dipole moment and therefore no pure rotational spectrum. Its lowest vibrational frequency is the bending mode at 667.3 cm^{-1}. This is near the maximum of the thermal spectrum. The line spacing and band spectrum of CO_2 is narrower than the H_2O spectrum.

The CO_2 greenhouse effect would be most unambiguous under a clear sky at low humidity. The temperature decrease at sunset shows that heat trapping is short lived. The local surface temperature is more dependent on air mass movement.

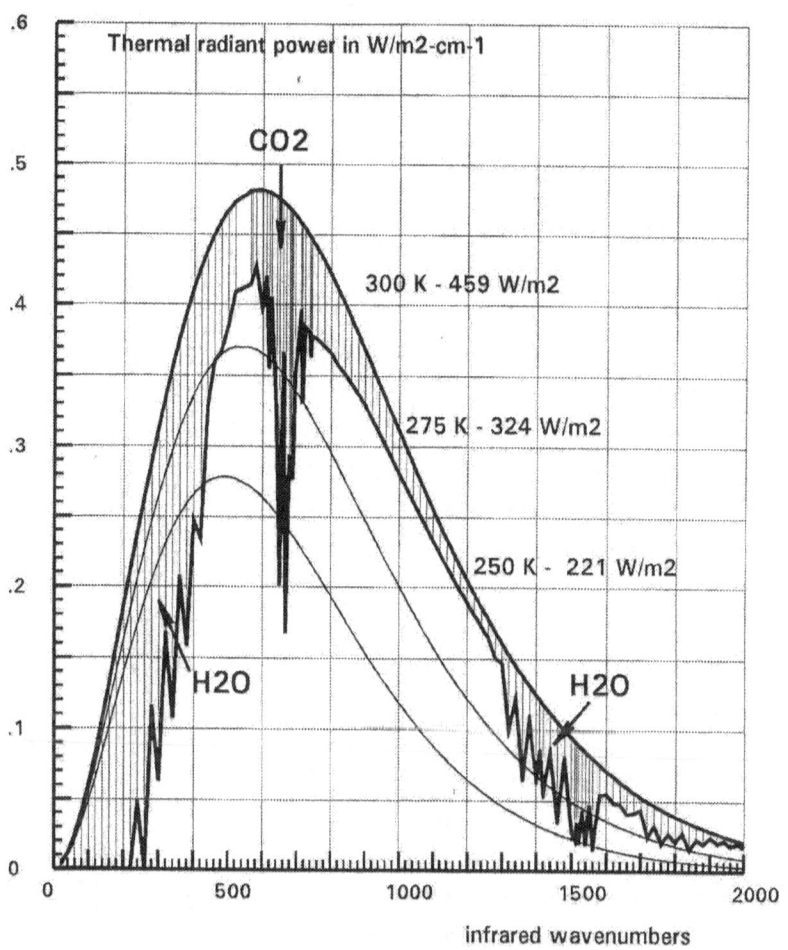

Fig. 6.2 Thermal emission compared with schematic greenhouse absorption bands

Fig. 6.3 shows the atmospheric carbon dioxide increase that is causing the alarm[14]. Since the site on Mauna Loa volcano in Hawaii is not in the path of prevailing wind from CO_2 sources it represents at least the northern hemisphere. Records from Antarctica confirm that the increase is global. The seasonal drop in CO_2 coincides with the Northern Hemisphere growing season.

Table 6.1 compares the 1.35 ppm mole percent average annual increase in atmospheric CO_2 with the fossil fuel components. Converting this to weight units give an average annual atmospheric CO_2 increase of about 3.2 Gt/y. The observed net average annual increase in atmospheric CO_2 is thus about half the total annual CO_2 increase from fossil fuel consumption.

Table 6.1 Fossil fuel consumption and atmospheric carbon

Coal consumption	2.4 Gt/y carbon
Petroleum consumption	2.5 Gt/y carbon
Natural gas consumption	1.6 Gt/y carbon
Total fossil fuel consumption	6.5 Gt/y carbon
Net atmosphere increase	3.2 Gt/y carbon

Atmospheric CO_2 absorption by plant growth and production by decomposition by plant and animal decay apparently reached a balance before the era of heavy fossil fuel use.

A natural question is why the CO_2 concentration does not dominate the Earth atmosphere as on Venus. The limestone cliffs that originated as sedimentary deposits in the oceans are obvious. They and other carbonate rock deposits account for over 99.5% of the carbon that exists outside the mantle. The oceans account for more than 90% of the remainder of the carbon. There is no possibility that atmospheric CO_2 can reverse this process.

The CO_2 increase by each year's fossil fuel consumption represents more or less the best million years of fossil fuel formation. This rate of CO_2 production is about double the rate at which the oceans can absorb it. This appears to be a more fruitful place to look for trouble than a greenhouse effect.

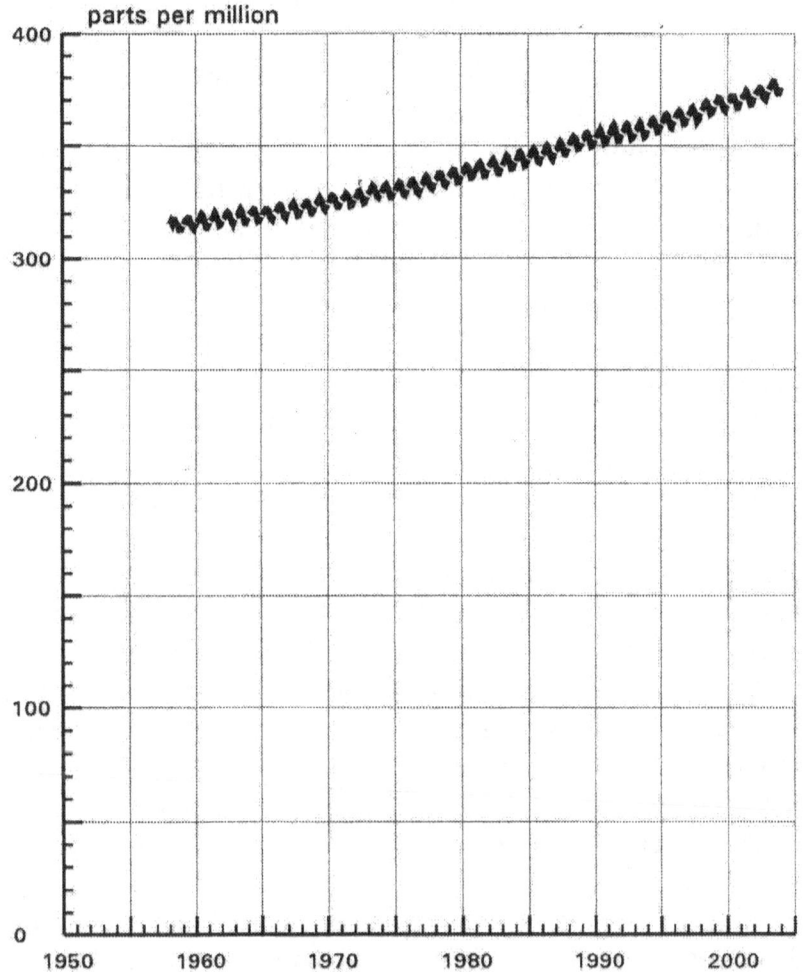

Fig. 6.3 Atmospheric carbon dioxide concentration

Data source: Carbon dioxide information data center,
Oak Ridge National Laboratory.

Heat Redistribution by the Atmosphere

Clouds are a highly visible indication that heat rising by convection plays a large role in climate dynamics. The *Nimbus-7 Earth Observation Satellite* observed the process that actually occurs from a polar orbit during 1979 and 1980[15].

Fig. 6.4 compares the energy from the sun with summaries of energy reflected from the Earth and emitted by the Earth as a function of latitude. Two sensors measured different aspects of the average radiation leaving the Earth. The uncertainties in the data are much larger than any net global warming or cooling that can be anticipated. They nevertheless demonstrate that the atmosphere plays a large part in moderating the Earth's climate.

A visible-ultraviolet sensor measured the radiant power of the *albedo* reflected during daylight. It accounts for about a third of the radiation from the Earth. The high reflectance at the poles is from cloud tops and polar ice caps. The high reflectance at lower latitudes is from deserts. Greater average cloud cover over temperate zones than over the tropics makes the albedo uniform. About two thirds of the radiation is reflected from water in clouds and one third from surface areas.

An infrared sensor measured *thermal radiant power* emitted from the Earth. Clear sky emission by the Earth's surface correlates closely with the ground temperature. The emission from cloud tops does not have a close correlation with the ground temperature.

The centers of high temperature circulation are near Indonesia, west of Peru, between Brazil and Africa, and west of Australia. Major high temperatures are also centered in the Sahara Desert, the Arabian Desert, and the desert region of central Australia.

Radiation at the poles is from polar ices caps and clouds. The average energy flow is from the tropics to higher latitudes.

The net difference between the solar input and the sum of the reflected albedo plus the infrared re-radiation shows that the energy lost at the equator balances the energy gained at the poles. This confirms that the general effect of the Earth's greenhouse gases is to transfer energy from the equator to the poles.

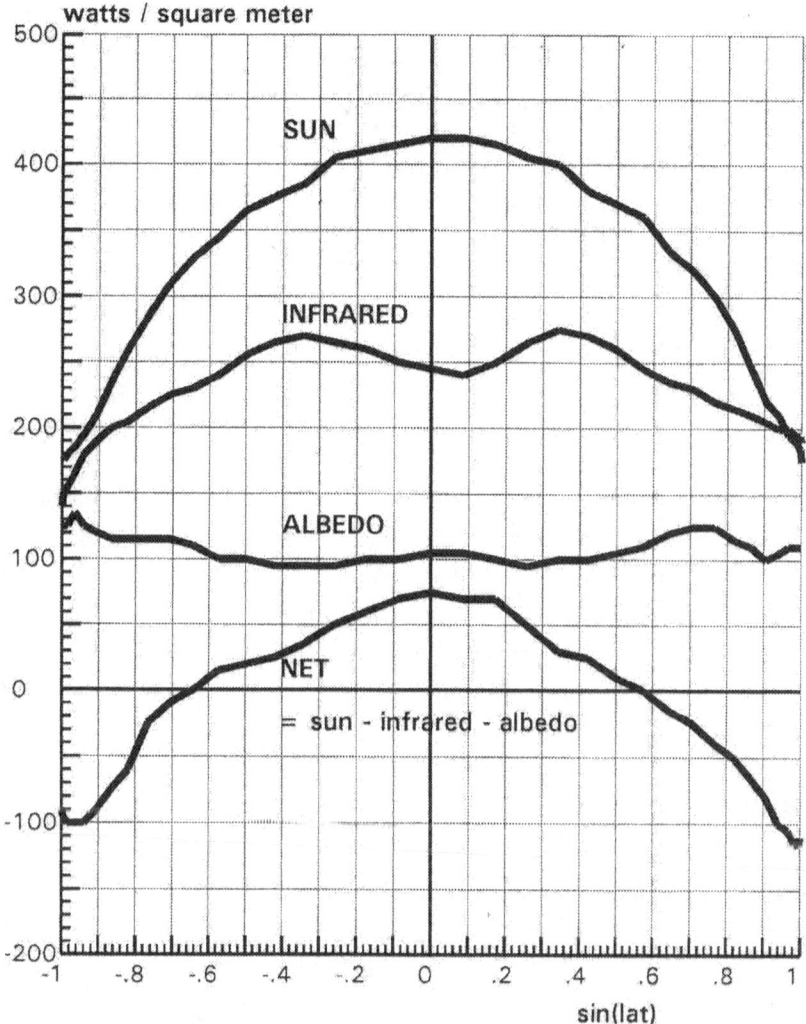

Fig. 6.4 Earth radiation balance as a function of latitude

Note: By plotting latitude as sin(lat) the areas under the curve
represent equal land areas.

Table 6.2 compares the capacity of atmospheric CO_2 and H_2O vapor to redistribute heat. They are both minor constituents of the atmosphere. They are both greenhouse gases. Their heat distribution and energy storage mechanisms are distinctive[16].

Table 6.2 Global greenhouse properties of H_2O and CO_2

	H_2O	CO_2
Mass in gas phase	54000 Gt	3066 Gt
Mass flux	1,020,000 Gt/yr	+11 Gt/yr
Specific heat	2443 kJ/kg	0.84 kJ/kg
Stored energy	36600 PWh	0.71 PWh
Energy flux	79 PW	1.8 PW
Energy holding period	19 days	8 hrs

Carbon dioxide is an inert, uniformly distributed component of the atmosphere. Its capacity to store heat depends on the local temperature and its specific heat which is relatively small. Any component of the local temperature that results from infrared absorption is mixed by convective air currents.

Water vapor is a dynamic component of the atmosphere with a wide range of concentrations. The annual rainfall is a measure of the average annual heat that water transports to the atmosphere. The energy transported to the atmosphere as heat of vaporization remains as kinetic energy of the cloud particles after it condenses as clouds. The temperature and energy of the clouds change with convection and expansion. Clouds contain most of the solar energy. They mix and transport it both vertically and horizontally. When the water eventually precipitates as liquid water, snow, or ice it forms heat reservoirs or sinks in the form of lakes, rivers, oceans, ice caps, and glaciers. The heat reservoirs make the daily and seasonal temperature limits much narrower than those on Mars or the Moon, where there is little or no atmosphere.

The energy stored by atmospheric water is up to 50,000 times greater than CO_2. The ratio of stored energy to heat flux shows that water can hold the energy long enough to produce climate change while CO_2 cannot. Both the specific heat and optical density of CO_2 are too small to have a large greenhouse effect on Earth. The effect of CO_2 on climate is not directly related to heat.

Global Warming

Global warming and receding ice sheets have been an aspect of northern climates and mountainous regions observed throughout recorded history. Many features of the Front Range of the Colorado Rocky Mountains are named glaciers in honor of pioneers of a century or so ago. The remnants of glaciers that were once thousands of feet deep are still obvious. By 50 years ago they had dwindled to permanent snow fields. Now they are only part of the annual snow cover that largely disappears by late summer. An even more obvious change of the past 50 years is that the clear view of the mountains from the eastern plain is gone. The once brilliant Milky Way is rarely visible. The smog that now obscures once clear skies is attributed to fossil fuel consumption. The increase in CO_2 does not contribute to this problem since it is a transparent gas.

Therelationshipbetweenglobalwarmingandglobaltemperature is not simple. Global temperature is not easy to define. More than 2000 sites worldwide have made temperature measurements for the past 125 years. Assigning temperatures to uniform elements of a global latitude-longitude grid partly compensates for the emphasis on populated land measurements. Further compensation for the well established urban heating effect must also apply. In this way the array of measured temperatures applies to a different array representative of an average global temperature[17]. Judicious mapping and extrapolation produces a temperature record for each grid element.

Since the grid is global the temperatures span the entire array of global climates. This range of temperatures is too broad to have an average value with a clear significance.

An average global temperature anomaly is a temperature difference that represents global warming or cooling. The reference temperatures for each grid element are a 30-year average over the years 1951 to 1980. The anomaly is the difference between the measured temperature and the reference temperature at each grid element. This temperature averaged over the entire grid is the monthly global temperature anomaly.

Fig, 6.5 shows the 5-year running average temperature anomaly. The correlation between the recent increase in both the temperature anomaly and CO_2 is widely cited as evidence that CO_2 is accelerating global temperature change.

Global temperature and global heat content are different concepts with much different values. Air, land, oceans, ice caps, and glaciers store different quantities of heat by different mechanisms at different rates.

The Moon is a useful comparison. Its surface is uniform. All but the top few meters is insulated from temperature changes. The temperature range is an enormous 260 ± 100 K but it tracks the heat content of the surface. Both the global temperature and heat content are nearly constant. An Earth with no oceans or atmosphere would be similar except for size.

Table 6.3 compares the heat in petawatt-hours per degree that can be stored by the Earth's major heat reservoirs in metric petatons, Pt. The heat content of the land surface and atmosphere are both minor and can change at an hourly rate. The oceans cover 70% of the surface. The top 1000 meters mixes on a yearly time scale. The deep ocean have most of the water. They mix much more slowly. Ice caps contain less than 3% of the water but can hold most of the labile heat. They form and disappear on a millennium time scale and longer. Circulation of the atmosphere and oceans drives the heat exchanges.

Table 6.3 Specific heat of different heat reservoirs

Atmosphere	5.1 Pt	1.0 PWh/K
Land surface	4.4 Pt	0.7 PWh/K
Surface ocean	332 Pt	387 PWh/K
Deep ocean	1331 Pt	1547 PWh/K
Ice caps and glaciers	28 Pt	47225 PWh

Global warming and global temperature are obviously difficult to define and measure to give a result with unambiguous meaning. The temperature history at a heat sink with a major influence on energy flows must be weighted heavily in the averages.

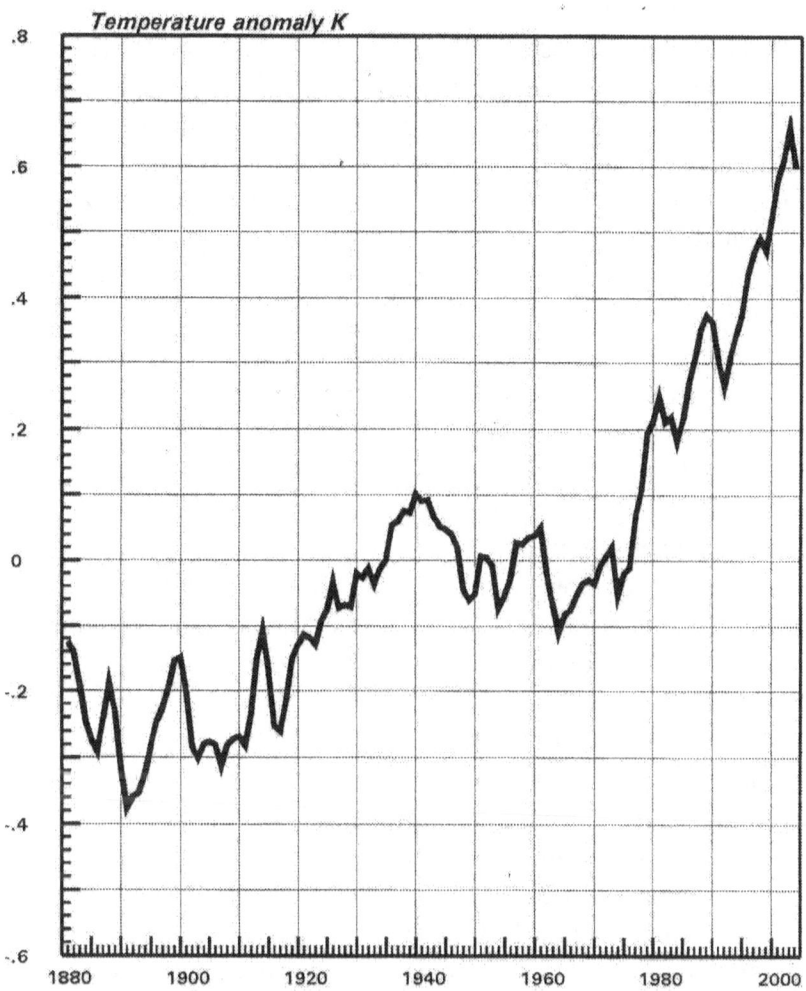

Fig. 6.5 Average global temperature anolmaly

Data source: Goddard Institute for Space Science

The history of receding glaciers and ice sheets is far longer than a century. Except for the 30 years between 1940 and 1970 the global temperature in Fig. 6.5 increased. It accelerated after 1970. The total increase is about 1°C over the century. This does not necessarily mean the acceleration is a prelude to catastrophe. It does not necessarily mean the increase in temperature is due to the increased CO_2 concentration. Geologists measure temperature over geologic time by analyzing cores through Earth sediments.

Fig. 6.6 is a 70,000-year history of the temperature of ice as it accumulated in Greenland. The deuterium/hydrogen isotope ratio increases with decreasing temperature[18]. The temperature cycle in upper regions of the core date the annual snowfall year-by-year. At greater depth the plasticity of the ice under pressure causes mixing that increases the uncertainty in detailed dates[19].

Low temperature near Greenland affects Atlantic Ocean currents. Freezing the salt water increases its density by increasing the salt concentration and by decreasing the temperature. This causes the surface flow from lower latitudes to sink to the bottom. The effect of the ocean on adjacent climates makes the temperatures of Greenland a reasonable surrogate for relative temperatures of the Northern hemisphere.

The first 40,000 years of the 70,000 year period was essentially an ice age punctuated by spikes of global warming. The magnitude of the warming and cooling episodes started to decrease about 20,000 years ago. This started a warming trend that reached its present value 10,000 years ago.

The most recent 10,000 years is an inter-glacial period of high temperature and exceptional stability. This period corresponds roughly to the origin and development of agriculture and organized human society until the present time. Average temperatures have decreased by 2-4 K with fluctuations of about ± 4 K at ± 0.013 K per year averaged over a century. From -70,000 to -10,000 years ago the snow deposition temperature varied in a 20 C range well below present temperatures. In the interglacial period the variation is less than 5 C. Changes during the past century are not remarkable in a longer-term context.

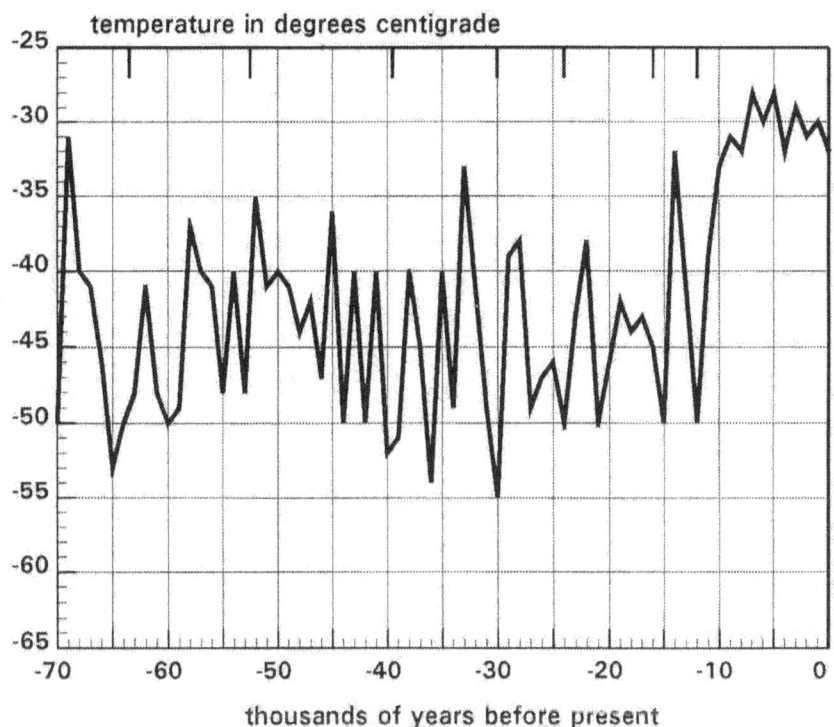

Fig. 6.6 70,000 year history of the temperature
of the Greenland Ice Sheet

By permission: Richard B. Alley, *The Two Mile Time Machine*, © Princeton
University Press

Circulation of the Atmosphere[20]

Horizontal air circulation is motion of the air mass as a part of the Earth plus motion of an independent mass to conserve momentum in interstellar space. The Earth's rotation does not change the direction of air moving along the equator. At higher latitude the momentum of the east-west component motion carries it further than the shorter radial path. The path curves away from the equator, an example of the *Coriolis Effect[21]*. It increases with increasing latitude from zero at the equator.

A low is a rising column of low density air with buoyancy from solar heating. The vector product of the Coriolis effect and air rising in a low is helical. Wind flow toward the base of a low is *cyclonic*, that is, counterclockwise in the northern hemisphere and clockwise in the southern hemisphere. Lows can be semi-permanent, seasonal, or mobile. They form over heated areas with access to hot, moist air. Mobile lows move according to the local pressure.

The tropopause is the top of the atmosphere where ozone and oxygen absorb solar heat and reverse the cooling. The air temperature in a low decreases with altitude by expansion. At the tropopause it cools further by thermal radiation. The temperature of the air and the altitude of the tropopause both decrease toward the poles.

A high is a column that flows downward due to the higher density of cold air at the top of the column. Wind flow toward the top of a high is *anti-cyclonic*, that is, clockwise in the northern hemisphere and counterclockwise in the southern hemisphere. Highs can also be semi-permanent, seasonal, or mobile. They form over areas with ice cover or over cold ocean areas that act as heat sinks.

Polar cold air masses form gigantic anti-cyclone highs at each pole. As they rotate they spin off smaller *mobile polar highs*, anti-cyclones that drift southward from the poles toward the tropics[22]. Each pole spins off a mobile polar high pressure air mass in some direction nearly every day on average. The topography of the land significantly influences the flow of cold air at the Earth's surface. The mobile polar highs are the origin of most weather patterns throughout the mid-latitudes.

Table 6.5 shows the gross global air circulation. The two extremes are the equator and the poles.

Table 6.5 Gross global air circulation features

Latitude Range	Horizontal flow type	Typical wind direction	Vertical flow type	Troposphere altitude, km
75-90 N	polar high	polar easterly	Polar cell	40
50-75 N	sub-polar low	westerly	Polar front	55
5-50 N	sub-tropical high	NE trade wind	Hadley cell	80
5S -15N	equatorial trough	doldrums	Convergence	00
5-50 S	sub-tropical high	SE trade wind	Hadley cell	80
50-75 S	sub-polar low	westerly	Polar front	55
75-90 S	polar high	polar easterly	Polar cell	40

Vertical air circulation is a pattern of cells formed by lows rising at low latitudes and highs descending at higher latitude. *Hadley Cells* are air rising from lows near the equator. The air flow away from the equation loses enough heat for part of the cold air to descend near 30° latitude. It turns partly westward to form easterly winds that converge toward the *doldrums* near the equator. Smaller and less well defined *polar front cells* transport heat from the mid-latitudes closer to the pole. Still smaller *polar cells* complete the transport of the coldest air to the poles.

Mid-latitude weather consists primarily of collisions between mobile polar highs and mobile sub-tropical lows moving in opposite directions. In the northern hemisphere mountain ranges extend from the pole like giant turbine blades to redirect mobile polar highs. Oceans surround Antarctica. Mobile polar high pressure masses in the southern hemisphere are more predictable.

Jet streams are high altitude winds that flow west near 30° latitude and east near 60° latitude. They compensate for the momentum of the easterly and westerly winds, respectively.

A significant role for CO_2 in the atmospheric processes that create weather and climate is not obvious. Its contribution to radiative cooling at the tropopause would produce global cooling. Atmospheric *aerosols* play a significant role in water vapor condensation. Fossil fuel produces atmospheric aerosols that are visible as smog. The smog does not involve CO_2.

Circulation of the Oceans[23]

Oceans store more than 90% of the carbon that is not in carbonate rock deposits. The ocean is essentially a 0.6 molar sodium chloride solution containing smaller amounts of 88 other elements. Ca^{++} and Mg^{++} ions give the solution a somewhat basic pH 8.1. This dissolves CO_2 as bicarbonate ions, HCO_3^-. The resulting carbon content of the oceans is 50 times greater than the atmosphere. The oceans are a reservoir for carbon as well as heat.

Solar radiation, temperature change, rainfall, influx from rivers, and ice formation create wide variations in the ocean surface conditions. Mixing is comparatively slow due to the size of the ocean. The *surface ocean* is the top 1000 meters with most of the variation. The *deep ocean* has colder, denser, saltier water that extends to an average depth of 3900 meters. It mixes with the surface ocean selectively and on a much longer time scale.

Land masses constrain the horizontal anticyclonic circulation. The *Gulf Stream* carries heat at 30° C from the equator along the Atlantic coast off North America to Greenland at -2° C in the Arctic winter. It continues south off Europe to the equator.

The *Japan Current* flows north from the equator at 28° C along the East Asian coast, across the North Pacific between Japan and Alaska at 3° C in the Arctic winter, then back to the equator along the U.S. west coast.

An ocean current circles Antarctica without direct heat from flow through warmer climates. This gives the Antarctic low humidity, arid climate, small annual snowfall, and lower temperature compared with the Arctic.

The *southern oscillation* creates periodic events known as *El Nino* and *La Nina* on the Pacific equator between Peru and Indonesia[24]. During an El Nino the normal easterly wind direction reverses blowing a plume of heated surface water toward Peru. This increases the cloud cover and rainfall over a wide area of South America. The temperature off Indonesia falls below normal as heated surface water blows eastward. A resulting upsurge of cold deep ocean water brings up plankton and attracts an abundance of fish.

A La Nina restores the normal flow direction and high temperature returns to Indonesia. The temperature of the waters off Peru falls below normal. The normal westerly surface wind intensifies, heats the surface, and gives the Western Pacific abnormally high temperatures, cloud cover, and rainfall. This brings cooler water to the surface and below normal rainfall over a large region of the Eastern Pacific.

CO_2 is more soluble in colder water. Up-welling cold ocean water releases it to the atmosphere. The $^{13}C/^{12}C$ isotope ratio distinguishes CO_2 of cold deep ocean water from the prevailing atmospheric CO_2. The low deep ocean temperature makes it the major CO_2 reservoir outside the mantle.

Evaporation in the tropics increases the salt concentration and density. During the flow northward fresh water from continental rivers partly counteracts the density increase. Density in the Polar Regions is the combined effect of increased salinity, low water temperature, and fresh water removal by ice. *Thermohaline circulation* starts when low temperature, high salinity surface oceans sink near the poles. The deep ocean flow follows the topography of the bottom of the North Atlantic from points between Greenland and Labrador and between Greenland and Iceland.

The path traced by isotope composition flows south from Greenland along the deepest part of the ocean. The shallow Drake Passage between South America and Antarctica forces the current eastward. It continues on the deepest path beyond Australia then turns northward. It ends east of Japan after about a millennium.

Fresh water input can disrupt the thermohaline flow. The marks along the top edge of Fig. 6.5 denote *Heinrich events* in the record of Atlantic Ocean deposits in high plains of a deep ocean plateau in the ridge off the coast of Spain[25, 26]. Interruptions in these cores are attributed to rock abraded by glaciers. Episodes of *ice rafting* in the glacier movement on the Iberian Peninsula occurred during the ice age. They coincide with periods of extreme low temperature in the Greenland ice core record.

Fig. 6.7 shows the Earth's major carbon reservoirs outside the mantle. The total carbon mass is 14.5 petatons. This is the cumulative total of the 11% CO_2 in volcanic gases. Unlike Venus, the early Earth formed liquid water oceans. Rainfall dissolved Ca^{++} and Mg^{++} ions from the rocks. In the oceans these combined with HCO_3^- ions as carbonate rock deposits. Some form of limestone deposits now sequester over 99.5% of the total carbon.

The Earth's oceans store carbon as bicarbonate ions, HCO_3^-. These ions contain more than 90% of the remaining carbon. The oceans contain 50 times as much carbon as the atmosphere. The surface oceans exchange CO_2 with the atmosphere. The known exchanges between the surface oceans and the deep oceans are the down-welling of cold dense water at the poles and the up-welling that occurs during events such as an El Nino. The exchange of global scale quantities is slow.

The net effect of fossil fuel on CO_2 is to produce 3.2 Gt/y more carbon than the surface ocean can absorb. The total quantity of carbon from fossil fuel is about double the quantity that accumulates in the atmosphere.

The questions about the effect of oceans are whether CO_2 accumulates in the atmosphere because the capacity of the ocean is too small, because the rate of transfer to the ocean is too slow, or because other factors are changing the capacity and/or rate. The concentration of HCO_3^- ions in the ocean decreases with increasing acidity of the ocean. The Ca^{++} and Mg^{++} ion concentration are the major constituents that regulate the acidity. Acid pollutants, such as sulfuric acid from high sulfur fuel, could increase the acidity (decrease the pH). This decreases the bicarbonate ion solubility.

The questions about climate change are how much of the present global warming is due to the increase in CO_2, how much is due to other human activity, and how much is part of the long term historical trend that is an inevitable consequence of planetary dynamics.

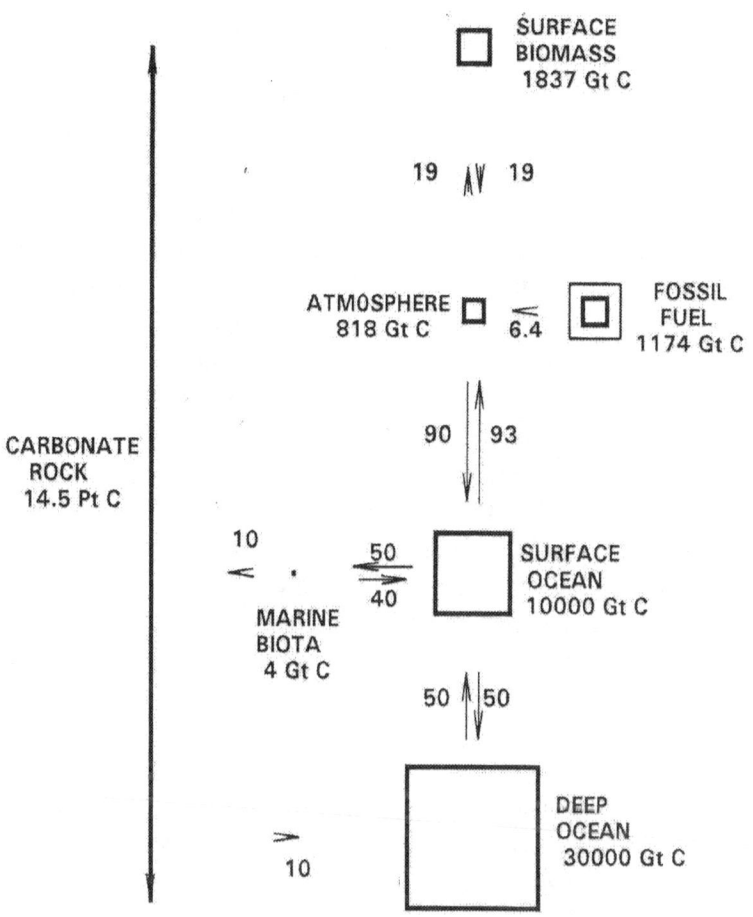

Fig. 6.7 Dynamics of carbon outside the mantle

Note: Squares represent cubic volumes proportional to carbom mass

Milankovic cycles

Milankovic cycles are the result of the Newtonian dynamics of rotating bodies that orbit the Sun.[27]. The Earth rotates with a 24-hour period.

The tilt of the earth's rotational axis with respect to the solar orbit axis is 23.45°. The tilt toward Sun and away from the Sun causes summer and winter. The tilt direction *precesses*. This changes the timing of the seasons with respect to the eccentricity of the solar orbit. This period is about 21,000 years. The tilt *nutates*, or wobbles. This changes the magnitude of the seasonal change. This period is about 45,000 years. The present tilt is near the center of a 23.23±1.23° range.

The eccentricity of the Earth's orbit changes with the alignment of the gravity of the Earth, Sun, and major planets. This has several periods of 100,000 years, or so. The radiant power is inversely proportional to the square of the distance from the Sun. The present eccentricity is nearly in phase with the tilt direction and partially cancels the effect of the tilt in the northern hemisphere. The closest approach to the Sun, the *perihelion* on January 2, nearly coincides with the maximum tilt away from the sun at the winter solstice on December 22.

Complex harmonic combinations of periods resemble those in geologic cores. The longest climate records are from cores through ocean deposits. One core in particular is from 3000-meter thick sediments on a high mountain plain on the Antarctic polar front of the Indian Ocean[28]. It records temperature and ocean life through the later ice ages. Four parameters cover a period of 450 million years[29].

Geological markers of known events give the date. The $^{18}O/^{16}O$ isotope ratio in the deposit gives the ocean surface temperature. Pollen species in the sediment indicate climate. Pollen grains are nearly indestructible and maintain the identity of the plant species for millions of years. Over 100 features of the temperature minima and maxima correlate with the Milankovich Theory after phase adjustments for best agreement with geological data. From 20 to 40 biological species correlate with expectations of climate at the prevailing temperature. This is the evidence that Milankovich cycles are an underlying cause of the ice ages.

Global Circulation Models

A global circulation model calculates the combined effect of all known processes. To create a model the first step is to divide latitude, longitude, altitude, and depth of the oceans into cells[30].

The model is a massive array of differential equations. The array contains an equation for the change in mass of each substance in each cell plus another array with an equation for the change in energy of each substance in each cell. Each equation contains a term for each process. The processes include chemical reactions, phase changes, and exchange with adjacent cells.

The solution to each equation is the change over an increment of time. The temperature, pressure, and composition of each cell evolve in time from an initial state with successive calculations.

The model is a caricature at best. The model covers an area of a half billion km^2. The area represents about 40 billion km^3 of atmosphere and a billion km^3 of ocean. The number of cells in the model, the details each cell covers, the accuracy of the independent initial conditions, and the time interval of each calculation are obvious compromises to be made.

The initial properties of each cell are valid only as far as they distinguish the forces for change in that particular cell at that particular time. The question is whether the same model is as valid for changes over a decade as for changes over many millennia. Weather models tend to lose most of the detailed agreement after only a few days[31].

Model calculations are only as significant as their validation by comparison with real data. They are an emerging science that is not yet validated by comparisons with the complexity, precision, or time scale consistent with climate change in the real world.

Notes and references for Chapter 6

1. The Kyoto Conference was organized under the International Energy Agency, an autonomous body within the framework of the Organization for Economic Cooperation and Development.

2. Beheron N. Kursunogly, Stephan Mintz, and Arnold Permutter, *Global Warming and Energy Policy*, Kluwer Academic Publishers, Dordrect, The Netherlands, 2001

3. Joyeeta Gupta and Michael Grubb, *Climate Change and European Leadership*, Kluwer Academic Publishers, Dordrect, The Netherlands, 2000

4. Satellites map 31 categories of Earth surface and its CO_2 consumption based on the growth characteristics. Trees, including most evergreens, typically consume 8-10 times more CO_2 than grains and grasses. J.S. Olson, J.A. Watts, and L.J. Allison, Major *World Ecosystem Complexes Ranked by Carbon in Live Vegetation*, Carbon Dioxide Information Analysis Center, Oak Ridge National Laboratory, 1985

5. The present rate of tree growth is the other aspect of the CO_2 unbalance to consider. Experiments show a significant correlation between the concentration of atmospheric CO_2 and its consumption by plant growth, nearly a direct correlation up to 40 percent for some enhancements. Leaves and needles, rather than stems, roots, or woody parts, consume the CO_2. The growth enhancement is greatest in young trees while their leaf growth is greatest. C.D. Idso and K.E. Idso, *Center for the Study of Carbon Dioxide and Global Change.* This is a private organization that has studied the effect of carbon dioxide on tree growth for 25 years. Their web site (www.co2science.org) includes a calculator that plots the temperature history at over 100 U.S. weather stations.

6. The relative pressure of CO_2 in paleo-atmospheres is inferred from the C^{13}/C^{12} carbon isotope ratio. According to this record the CO_2 pressure was 10-20 times its present value 450 million years ago. During the carboniferous era 350 to 270 million years ago it decayed to roughly its present value. In about 20 million years, to four times its present value. It has since decayed to the pre-industrial value of 300 parts per million. The initial decay is attributed to erosion and eventual formation of carbonate rock. Robert A. Berner, *The Rise of Plants and Their Effect on Weathering and Atmospheric CO_2*, Science, **276**, 544-546, 1997

7. Radiation temperature is the temperature at which an ideal black body would emit the same thermal radiant power as the radiation in question. The maximum solar radiation temperature on Earth is 364 K. This corresponds to the solar constant, 1.37 kW/m^2.

8. Mathew P.Thekaeka, NASA Technical Report R351 Measurements of the solar spectral radiant power to the top of the Earth's atmosphere made by a NASA high altitude research aircraft

9. The clear sky solar spectral radiant power measured at the Earth's surface. National Renewable Laboratory, Golden, Colorado

10. E. Friis-Christiansen and K.Lassen, *Length of the Solar Cycle; an indicator of Solar Activity Closely Associated with Climate*, Science, 254, 698 1991

11. The sun is brighter when the duration of the magnetic event is shorter. Sallie Baliunas and Willie Soon, Astrophysical Journal, **450**, 896 1995

12. R.W. Spencer and J.R. Christy, *Precise Monitoring of Global Temperature Trends from Satellites*, R. Jastrow, W. Nierenberg, and F. Seitz, *Scientific Perspectives on the Greenhouse Problem*, George C. Marshall Inst., Jameson Books, Ottawa IL 1990

13. The best indication of radiation trapping would be line reversal and other changes in the shape of CO_2 emission observed from above the atmosphere. The optical thickness of the atmosphere does not appear to be sufficient for this to occur. A succession of absorption and re-emission allows preferential transmission by the wings of the spectrum to become dominant. Bernard E. Douda and Edward J. Bair, *Radiative Transfer Model of a Pyrotechnic Flame*, J. Quant. Spectroscopy and Radiative Transfer, 14, 1091-1105,1974

14. The *Carbon Dioxide Information Analysis Center* of the Oak Ridge National Laboratory is a lead agency disseminating data related to global climate change.

15. H. Lee Kyle, et.al, *Atlas of the Earth's Radiation Budget*, NASA, Scientific and Technical Information Service, 1980. The NASA report on mapping by the Nimbus 7 research satellite of radiation emitted and reflected by the Earth

16. The CO_2 masses are obtained from the mole fraction data using the total mass of the atmosphere. The H_2O mass assumes 30% average relative humidity at 290 K. The H_2O mass flux is obtained using 2 m average annual rainfall over the Earth surface and the heat of vaporization, 2443 kJ/kg.

17. The Goddard Institute for Space Science global average temperature is from 2000 of the 8000 stations of the Global Historical Climatology Network to represent the past 125 years. The temperature of each 7870 km² element of the 64800 element grid obviously requires extrapolation.

18. Richard B.Alley, *The Two-Mile Time Machine*, Princeton University Press, 2000. A popular account of details of the acquisition and interpretation of an ice core at the GISP-2 site in central Greenland

19. K.M.Cuffey, and Clow, *Temperature, accumulation, and ice sheet elevation in central Greenland through the last de-glacial transition*, J. Geophysics. Res., 102(C12), 26, 383, 1997 Surface temperatures from Hydrogen/Deuterium isotope ratios in the 10,000 foot length of core ice obtained at the GISP-2 site.

20. Adrian Gordon, Grace Warwick, Peter Schwerdtfeger, and Roland Byron-Scott, *Dynamic Meteorology, a Basic Course*, John Wiley & Sons, New York, 1998, Vector algebra of motion in the gravitational field of the rotating Earth

21. The Coriolis Effect is the apparent force required to conserve momentum in a rotating frame of reference. Vertical flow due to changes in buoyancy of the atmosphere depends on the gravitational force toward the center of the Earth and the centripetal force toward the axis of rotation. In the middle latitudes the difference in direction of these forces interacts with the rotation of the Earth producing spiral cyclonic motion in rising air currents and anti-cyclonic motion in downward air currents. Cyclonic motion is counterclockwise in the northern hemisphere and clockwise in the southern hemisphere.

22. Marcel Leroux, *Dynamic Analysis of Weather and Climate*, John Wiley & Sons, New York, 1998 Aspects of weather and climate associated with mobile high pressure anti-cyclones

23. Dan Seidov, Bernard J. Haupt, and Mark Maslin, Eds., *The Oceans and Rapid Climate Change Past, Present, and Future*, American Geophysical Union, Washington, D.C., 2001

24. Ulrich Siegenthaler, *El Nino and Atmospheric Carbon Dioxide*, Nature, 345, 295, 1990.

25. Helmut Heinrich, *Origin of Ice Rafting in the Northeast Atlantic Ocean During the Past 130,000 Years*, Quaternary Research, **29**, 142-152, 1988

26. The increased ice rafting and onset of periods of extreme cold is attributed to the lower salinity of the surface ocean. This correlation is similar to a related, but smaller, more frequent pattern, known as *Dansgaard-Oechger events*. W. Dansgaard, et.al. *Evidence for General Instability of Past Climate from a 250,000-Year Ice Core Record,* Nature, **364**, 218-220, 1993; H. Oeschger, et.al. in *Climate Processes and Climate Sensitivity*, F. Hansen, ed., *American Geophysics Union Monograph 29*, Washington, D.C. 1984

27. The original theory of Milankovic has been reinterpreted and refined, K. Milankovic, Serb. Akad. Beoorg. Spc. Pub., **132**, 1941, N.J. Shackleton, *The Phanerozoic Time Scale*, Geological Soc., London, 1971

28. J.D. Hays, John Imbrie, N.J. Shackleton, *Variation in the Earth's Orbit: Pacemaker of the Ice Ages, Science*, **194**, 1121, 1976

29. R.C.L. Wilson, S.A. Drury, and J.L. Chapman, *The Great Ice Age*, Routledge, London, 1999

30. In the Euler model fluid flows through a set of cells fixed in space. In a Lagrange model cells with a fixed quantity move through space.

31. In the technical sense models are chaotic when negligible changes in the initial conditions cause unpredictable and radically different results. Force combinations that produce chaos are called *strange attractors*. John L. Casti devotes a chapter to weather prediction and modeling with examples of chaos theory. *Searching for Certainty*, William Murrow, 1991

Chapter 7
Energy Transformations

Thermal energy sources now produce most of the low cost electric power in spite of the Rube Goldberg character of the systems. A heat source first heats a gas to a high temperature and pressure. The hot gas expands against a piston or the blades of a turbine. This transfers kinetic energy to the output shaft. The shaft then drives a generator that produces electric power.

Fossil fuels now supply cheap reliable heat for electric power. The demand threatens to overtake the supply. Hydroelectric power is the only significant low cost energy source not based on a heat engine. To evaluate the alternatives it is useful to understand more broadly what energy is, how different forms are related, and how much is convertible to electric power.

Kinetic Energy

Kinetic energy is the energy of mass due to its velocity. This includes motions of large masses with observable consequences such as wind or rivers. It also includes the internal motions of atoms and molecules that make up a *system* with larger mass.

Translational motions of individual atoms and molecules in a particular direction are the overall translation of the system that contains them. Translational motions that are in random directions are a part of the heat that gives the overall mass a temperature.

Internal motions of the atoms of each molecule are also part of the heat of the system. These include *rotational* motions of the entire molecule, *vibrational* motions that stretch the bonds or change the bond angles of a molecule, or *electronic* motions of the electrons that create the bonds between atoms.

The internal energy of each molecule is *quantized*. Harmonics limit each mode of motion to specific rotational, vibrational, and electronic energy states. These are *allowed states* of the molecule. All other molecular energies are *forbidden states*.

Transitions between energy states of molecules may occur due to collisions between molecules. They may also occur by emitting or absorbing radiation. Radiative transitions are restricted to photons having energy that matches the energy difference between allowed states of the molecule. Photons of radiation have exactly the energy that matches the energy between two allowed states. The frequency of the photon is proportional to its energy and inversely proportional to the wavelength. Transitions between rotational states are generally at microwave frequencies. Vibrational transitions are mostly in the infrared. Electronic transitions are at ultraviolet frequencies and higher. *Radiant power* depends on both the energy per photon and the number of photons per second. *Spectral radiant power* specifies both the photon energy and rate .

The state of a system is characterized by *state variables*, macroscopic properties like temperature, pressure, and composition. They define the state of the system without regard to how it is formed. All systems in the same state have the same behavior. Although many state variables describe different features of a system only three variables for each substance are independent. Thermodynamics provides a way to calculate all state variables given any three[1].

In most molecules many allowed energy states are populated at normal temperature. Statistical thermodynamics derives formulas for the fraction of molecules in each allowed state. Precision is possible because the number of molecules in a system is very much larger than the number of allowed states. Under most conditions thermodynamic quantities are predictable.

Thermal equilibrium is the most important condition for predictability. The temperature, pressure, and composition of a system must be uniform for their value or the state of the system to be defined. In the absence of significant external forces the kinetic energy of most systems tends to move them toward uniformity and equilibrium. The laws of thermodynamics apply to systems that are close to thermal equilibrium.

The state of a system depends on the variables at equilibrium whether it is before a change or following a change. It does not depend on what caused the process to occur or what conditions exist during the process.

Potential Energy

Potential energy is energy due to the position of a system in a force field. The energy of a water reservoir, for example, is the product of the gravitational force, the mass times the gravitational constant, times the height or *displacement* above some reference level, that is, a force times a displacement.

Table 7.1 relates changes in the energy, dE, to a variety of analogs of force that might change it[2]. In general a change in energy changes all of the variables that describe the state of the system regardless what causes the change.

Table 7.1 Variables describing different energy sources

mechanical energy, f dx	- force x displacement
hydraulic energy, P dV	- pressure x change in volume
surface energy, γ dA	- surface tension x change in area
electrical energy, V dq	- voltage x change in charge
radiant energy, ε dn	- photon energy x number of photons
chemical energy, μ dn	- chemical potential x moles reacted[2]
thermal energy, T dS	- temperature x change in entropy

The force and displacement of water in a reservoir fit the definitions of Newtonian mechanics. Analogous *conjugate pairs* of variables describe the broader range of possible energies that might generate electric power. Like force times displacement they each have energy dimensions, kg m^2/s^2.

The intensive variable analogous to the force causing change describes the quality of the force. The frequency of the radiant energy, the magnitude of the voltage, the molar free energy of the chemical potential, or the temperature of the heat source all determine the effectiveness of a force in comparison with alternatives. A higher quality force generally means higher efficiency energy conversion.

The extensive variable analogous to displacement that determines the potential energy change is the quantitative aspect of force causing the change. It is generally in units per mole. A lower quality force requires greater displacement.

Work, Heat, and Entropy

Work, δw, is the energy of a particular kind of potential energy change. It is not a state variable because the state of a system is independent of the process that formed it. Since different kinds of work can cause the same change in the energy state dE, the change of the state, is different from δw, the energy of a particular process.

Heat, δq = T dS, is distinguished from work by the special properties of entropy and temperature. In other forms of potential energy the extensive variable makes intuitive sense as an analog of a displacement. Although temperature is a plausible thermal force the intuitive sense of a change of entropy, dS, as the displacement caused by temperature is not immediately evident.

Entropy is the extensive variable or displacement that defines a change in the energy state of a system due to temperature, dE = T dS. Adding or subtracting increments of heat, δq, in steps small enough to maintain thermal equilibrium changes the entropy of a system, dS = δq/T. Entropy differs from other extensive variables in many respects. Unlike mass, energy, and momentum, it is conserved only in systems which remain in thermal equilibrium.

The most probable energy distribution is the state with the largest number of different combinations of molecular states with the given total energy. This is also the thermal equilibrium state. As a property of a system at equilibrium, entropy is a *state variable*. The equilibrium distribution has the largest number of ways of distributing the energy among the allowed states. Since this is the most random distribution, entropy is associated with randomness.

A reversible process is one that is slow enough for the temperature to be well defined at each stage. The entropy change in a reversible process is dS = δq/T.

An irreversible process is usually too fast to equilibrate continuously. The existing combination of states is less than the maximum. In a process which departs from equilibrium the intermediate stages are undefined, dS > δq/T. Since the intermediate states are undefined the process is irreversible.

To picture reversible and irreversible processes consider a system containing air at atmospheric pressure in a cylinder closed by a piston. First double the pressure on the piston slowly. The gas pressure matches the pressure on the piston. The system remains arbitrarily close to equilibrium and the process is reversible.

Next latch the piston in place and add enough force to double the pressure. This defines the final state. On suddenly releasing the latch the external pressure remains double the pressure of the gas until the piston moves. Between the initial state and the final state the gas passes through turbulent states with pressures and temperatures that are not uniform and are thus undefined.

Calculations of otherwise identical reversible and irreversible processes yield the following generalizations. The work a system performs when it expands is smaller if the process is irreversible. The external work needed to compress a system is greater if the process is irreversible. Natural processes are spontaneous and irreversible to some degree. The entropy of the system plus the external surroundings increases regardless of the direction of the energy flow. For a cyclic process that returns the system to the original state the entropy of the system increases if any step is irreversible. Otherwise it is zero. Entropy increases as the energy of a system increases because a larger number of populated states have more ways to distribute the energy. Entropy approaches zero as the temperature approaches absolute zero, 0 K.

Absolute zero temperature, zero degrees Kelvin, appears to be the limiting temperature of outer space. The black body temperature of radio frequency radiation from dark portions of outer space is about 3 K. This is consistent with expansion of the universe over the 15 billion years since the Big Bang. All bodies in the universe lose energy by radiation toward outer space.

A standard reference state near the local ambient conditions is a more practical way to describe thermodynamic processes and substances[3, 4]. This state is typically atmospheric pressure and a temperature of 298 K or 25 C near ambient conditions. To perform useful work an intensive variable changes spontaneously from well above the standard state to a value nearer the standard state.

Work by Expanding Gas

Fig. 7.1 shows the P-V-T states for 1 mole of a gas that obeys the ideal gas law. A high temperature and pressure state is marked by the circle at upper left. The standard state is marked by a circle at lower right. The lighter curves are *isotherms*, the family of states that exist at particular temperatures.

Spontaneous expansion of a gas at high temperature and pressure generally produces a combination of heat and hydraulic work. An *isothermal expansion*, along an isotherm, would continuously exchange enough heat to remain at the temperature of surroundings at that temperature. An *adiabatic expansion* exchanges no heat. It would do hydraulic work on surroundings such as a piston or turbine blade. The heavy curve in the figure shows the temperature change by an adiabatic expansion.

The properties of the apparatus determine the path of an actual expansion. One lower limit of the expansion is set by the volume of the 1 mole of gas that expands. In an adiabatic limit the gas reaches the low temperature limit well before it has expanded to the volume of the standard state. In the isothermal limit the gas must expand far beyond the standard state to a volume at reduced pressure. The heavy part of the adiabatic curve is for an apparatus with a volume compression ratio of 5.

The high temperature reservoir that applies to expansion from this particular high energy state contains 1 mole of ideal gas. The gas can produce work by flowing spontaneously if there is a volume at low pressure that can receive it.

The low temperature reservoir is the mass of gas after the expansion. The work includes changes that produce the high temperature reservoir as well as those which produce useful work. If the mass is a flow rate the work is in units of power.

The net useful work of the expansion is the shaded area under the curve. It is the sum of PdV energy increments. The initial part of the curve does by far the greatest part of the work. There are diminishing returns as the gas approaches conditions near the standard state.

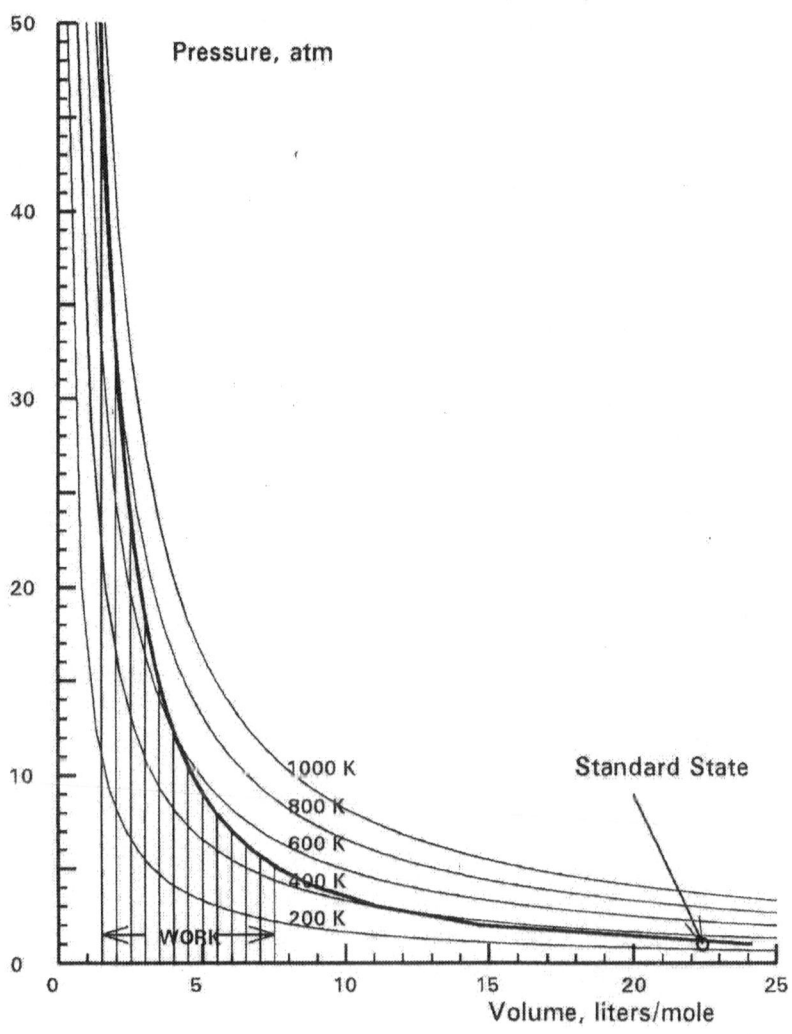

Fig. 7.1 Work done by expanding gas

Work in a Continuous Cycle

Fig. 7.2 shows the *Carnot engine* **cycle**, a reversible cycle that produces net useful work. Table 7.2 describes the processes. To understand the cycle, compare each step in the table with the corresponding step in each diagram.

Table 7.2 Steps in the cycle of a Carnot engine.

1→2 The system expands isothermally. This does external work and absorbs external heat that maintains the initial temperature.

2→3 The system expands adiabatically. This does external work without exchanging heat from the surroundings.

3→4 External work compresses the system isothermally. It also transfers heat to the surroundings to maintain the low temperature.

4→1 External work compresses the system adiabatically raising the system to its initial high temperature without exchanging heat.

The object of any heat engine design is to devise a series of processes that returns the gas to its initial state by a path that encloses the largest possible area of a PVT surface. The virtue of the Carnot engine is the simplicity of the cycle. It is more an illustration of principles than a practical heat engine. There are mathematical proofs that a Carnot engine produces the maximum work that is possible between two given temperature states.

The PVT diagram at the bottom shows the PdV hydraulic work, the area the cycle encloses. Since the complete cycle returns the gas to its initial state, the net energy change for the cycle is zero. Therefore the sum of the net heat exchanged and the net work done must also be zero. The net work enclosed by the PVT cycle equals the net heat that produces the work.

The temperature-entropy TS diagram at the top is the corresponding TdS heat. The plot is a rectangle. The sides are adiabatic changes in temperature without exchanging heat. The top and bottom are isothermal processes that exchange heat with the surroundings without changing the temperature. Compressing the gas isothermally generates heat. The area of the TS heat cycle encloses the heat that is converted to work.

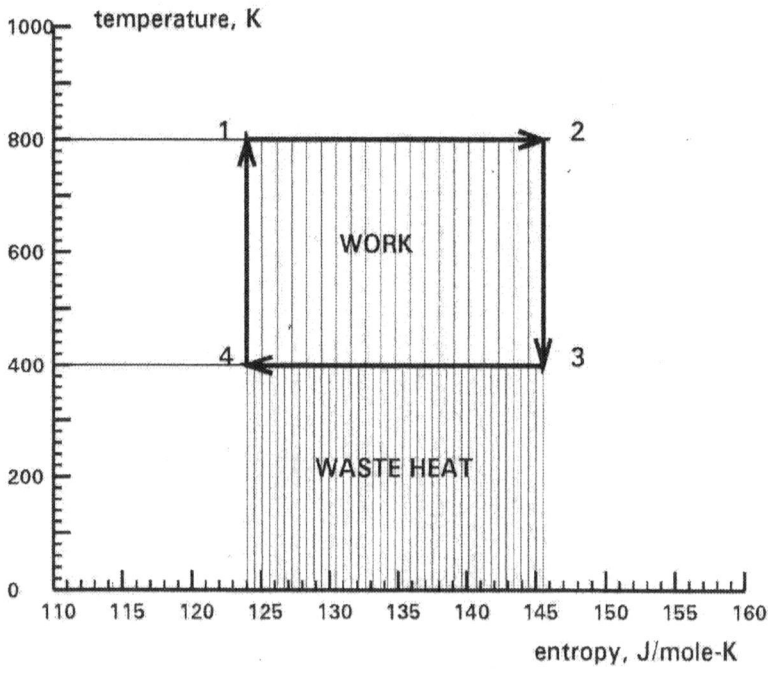

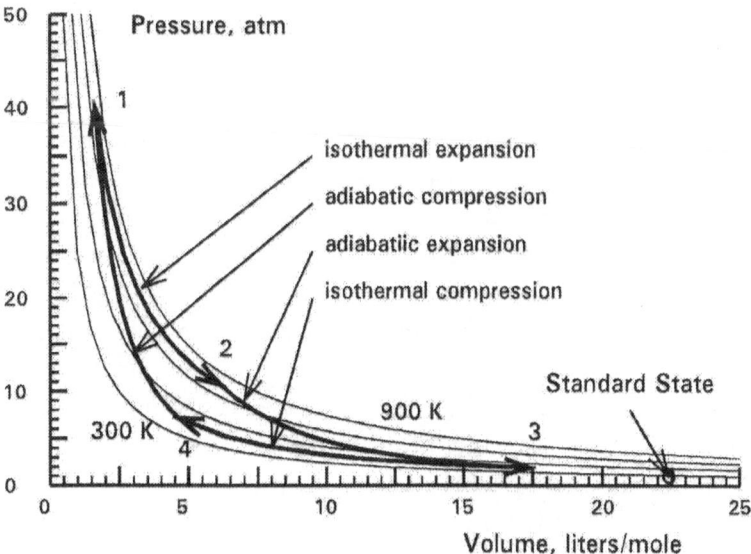

Fig. 7.2 PVT work and TS heat in a Carnot cycle

The efficiency of a heat engine is the ratio of the net work, w_{net}, produced by the gas to the heat, q_h, added to the gas at the high temperature, T_h. The efficiency of the Carnot engine is evident by inspection of the T-S diagram. The work, w_{net}, is equal to the difference between the two shaded areas. The heat added at the high temperature, q_h, is the sum of the shaded areas. The efficiency in this example is 400/800 or 50%. The efficiency of practical heat engines is more typically 33%.

The complete cycle returns the system to the initial state. The net change in the state variables, energy and entropy, is zero and $w_{net} = q_h - q_c$. Since the entropy change for a complete cycle is zero, $q_h/T_h = q_c/T$. The efficiency can be expressed in three equivalent ways. The efficiency of an ideal Carnot engine depends only on the temperatures of the hot and cold reservoirs.

$$\varepsilon \equiv - w_{net} / q_h$$

$$= (q_h - q_c) / q_h$$

$$= (T_h - T_c) / T_h$$

The value of heat in relation to electric power is the entropy change in relation to the absolute zero of temperature. The change in entropy in the real world is more often relative to the entropy at a realistic temperature of a low temperature reservoir. The heat and change in entropy in the real world is therefore only a fraction of the theoretical limit. In the Carnot engine the exhaust heat is most tangible during the isothermal compression.

Murphy's Laws of thermodynamics explain the dilemma. (1) A heat engine with perfect efficiency requires a cold reservoir at absolute zero temperature. (2) The cold reservoir must absorb a large quantity of heat. (3) A cold reservoir at absolute zero cannot absorb a large quantity of heat and remain cold.

Removing heat from engines that operate in a closed cycle is not a trivial problem. Some power plants do this by transferring the heat to water which escapes as steam. A power plant producing 1 GW of electrical power also produces 2 GW of low level heat that must be disposed of in other ways. It takes a water supply of about 3000 tons per hour, or 1.5 million gallons per hour, to produce enough steam to carry 2 GW of heat as the heat of vaporization of water.

U.S. surveillance satellites could easily recognize the Russian nuclear weapon facilities buried deep inside a mountain at Krasnoyarsk Siberia. Upstream from the mountain the Yenisei River is frozen and covered with snow. Downstream from the mountain it flows freely.

In a practical heat engine the processes that produce work are generally adiabatic. They occur at a rate that is too fast to exchange significant heat with the surroundings. This is inherent in both the rates of different kinds of energy exchange processes and in the mass flow rate of gases through the engine components that generate power. Adiabatic processes can have almost unlimited speed. A reversible adiabatic process must be slower but the processes of internal equilibration are usually fast in comparison with exchanges with the surroundings. Exchanging heat with the surroundings is slow. Exchanging heat between gases and its surroundings is particularly slow.

Heat engines that operate at a low speed are inefficient for reasons beyond the thermodynamic efficiency of gases. The power from a heat engine is proportional to the mass flow rate of the gas. The mass flow rate of gas through an engine is the same at each point in the cycle. However where heat exchange is necessary it is easier to design an efficient heat exchanger with a large surface area than to design an efficient engine that is large enough to operate slowly.

Brayton Cycle Gas Turbine Engine

Fig. 7.3 is a block diagram of a Brayton cycle engine. The Brayton cycle describes the basic characteristics of most gas turbine engines. The step numbers start with the physical starting point, the air input to the turbine.

Table 7.3 details the individual steps. The separate input and exhaust make this an open cycle Brayton engine. In a closed cycle Brayton engine the input and output are connected and the working fluid is sealed in the system. An energy source such as solar power replaces combustion in a closed cycle.

Turbine blades that compress the gas input, other turbine blades that produce power, and the electric power generator are components that rotate on the same shaft. The compressor uses a portion of the power from the power turbine. The adiabatic expansion in the power turbine reduces the pressure driving the gas by a compression ratio of 2.6 while the gas is still 750 K. The combustion products emerge from the turbine at a temperature much higher than ambient.

The recouperator uses the excess heat of the gas leaving the turbine to preheat the air input to the combustion chamber. The flow and temperature change in step 2→3 is equal and opposite to that of the compressor output in step 5→6.

Table 7.3 Steps in the Brayton engine cycle

1→2	Adiabatic compression raises the temperature and pressure of the ambient air intake.
2→3	The low pressure (power output) side of the recouperator transfers excess heat from the power turbine output to the air input.
3→4	Air-fuel combustion heats the gas working fluid to its maximum temperature and pressure.
4→5	Adiabatic expansion delivers power to the engine shaft and reduces the temperature.
5→6	The low pressure (power output) side of the recouperator transfers excess heat from the power turbine output to the air input.
6→1	The recouperator exhausts the output to the ambient surroundings.

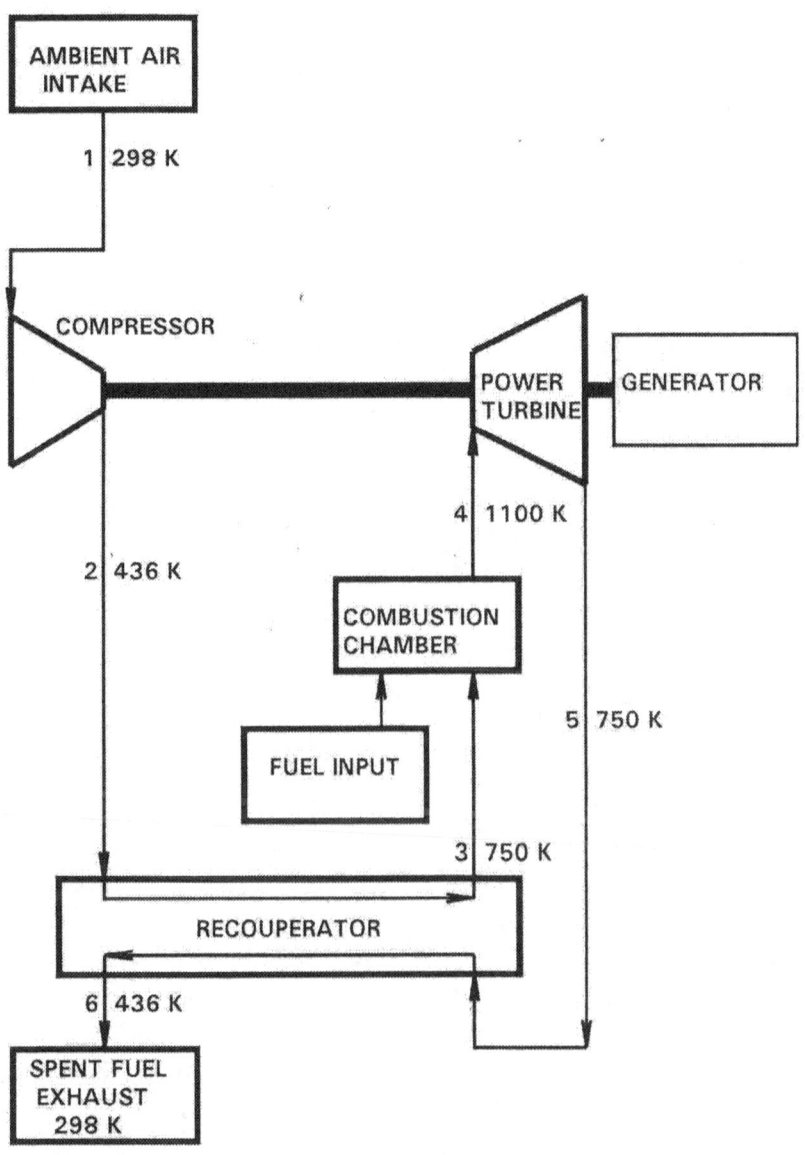

Fig. 7.3 Brayton gas turbine block diagram

Fig. 7.4 shows the thermodynamics of the engine in the block diagram. The six numbered steps correspond to physical states of gases flowing through the engine at a given mass flow rate. The high temperature and pressure difference of the combustion products creates the gas flow that drives the turbine.

The exhaust heat sink is the ambient atmosphere at 298 K. Adiabatic compression heats the input air to 436 K. The recouperator and combustion heat it further to 1100 K. The high and low pressure gas flow in the recouperator is in opposite directions. The temperature drop in the low temperature side is the mirror image of the temperature rise in the high pressure side. The heat from the source, which is nominally equal to the energy delivered by the turbine, changes the temperature from 750K to the maximum temperature, 1100 K.

The shaded areas under the T-S diagram show that half of the total heat provided by the high temperature source is discarded as waste heat, and half is converted to useful work. The shaded areas under the PVT curves show that half the total work done by the gas must be used to compress it back to the initial state pressure. The gas turbine performance in this example is high due to the high temperature.

The efficiency of the Brayton engine increases with increasing *compression ratio*. This ratio also increases the fraction of the heat input that must circulate in the recouperator. A high temperature is selected for high efficiency. The relatively low engine pressure is consistent with the strength of materials at high temperature. This requires a combination of high flow rates and/or a large engine volume. The high and low temperature limits and a compression ratio of 2.6 give the heat circulation shown.

In peaking power plants the heat is supplied by combustion of natural gas. As described further in Chapter 10, it can also be supplied by solar energy. Some gas turbine installations function as combined cycle operations by replacing the recouperator with a steam turbine. This improves the efficiency at the cost of some of the start-up flexibility.

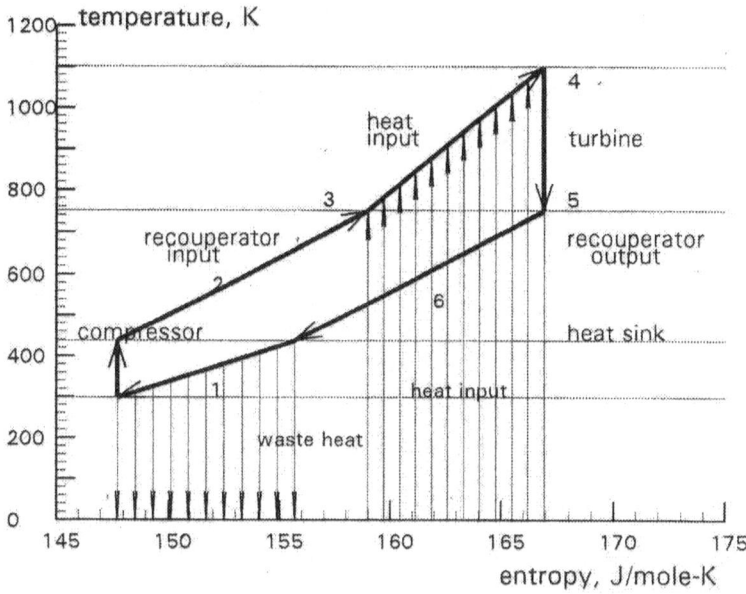

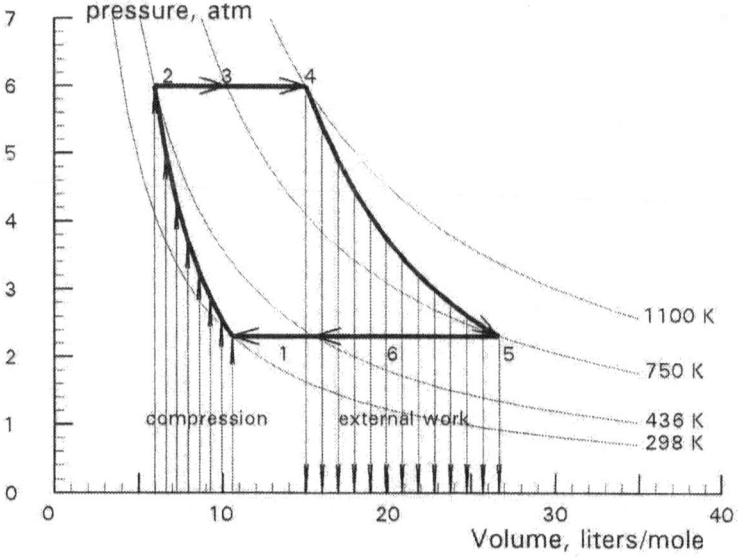

Fig. 7.4 T-S and P-V-T plots of a Brayton cycle

Electric Power from Steam

Fig. 7.5 shows PVT states of water that are relevant to a steam engine. A gas that condenses to liquid at a little above the ambient temperature can simplify the compression part of an engine cycle. The PVT states of steam in the region where it can condense to liquid are the key to a steam engine.

In comparison with an ideal gas this figure shows the states that have smaller molar volume and higher pressure. The range of this figure includes liquid-water mixtures as well as steam. This expands the left side of the view in Fig. 7.1 toward higher pressure.

The steam saturation curve is the heavy curve. These are states of steam that are saturated with water and ready to condense[5]. In compressing the ideal gas along the 600 K isotherm that is emphasized water behaves as an ideal gas up to the steam saturation curve. At the saturation curve the steam begins to condense as a water-steam mixture. With further compression the mixture contains more water and less steam in proportion to the distance toward zero molar volume. The isotherm becomes a horizontal step at constant pressure. The states below the saturation curve are mixtures of steam and water.

At the molar volume of liquid water the system becomes essentially incompressible. Further attempt to decrease the volume causes a nearly vertical pressure increase at the low compressibility of liquid water. With increasing temperature the volume range where steam-water mixtures are possible decreases and the horizontal steps become shorter.

The critical point is the unique highest temperature, highest pressure, and lowest molar volume state where water-steam mixtures co-exist[6]. At temperatures above the critical temperature the system is *super critical steam*. This is the state of water in the high temperature reservoir of a high performance steam turbine engine.

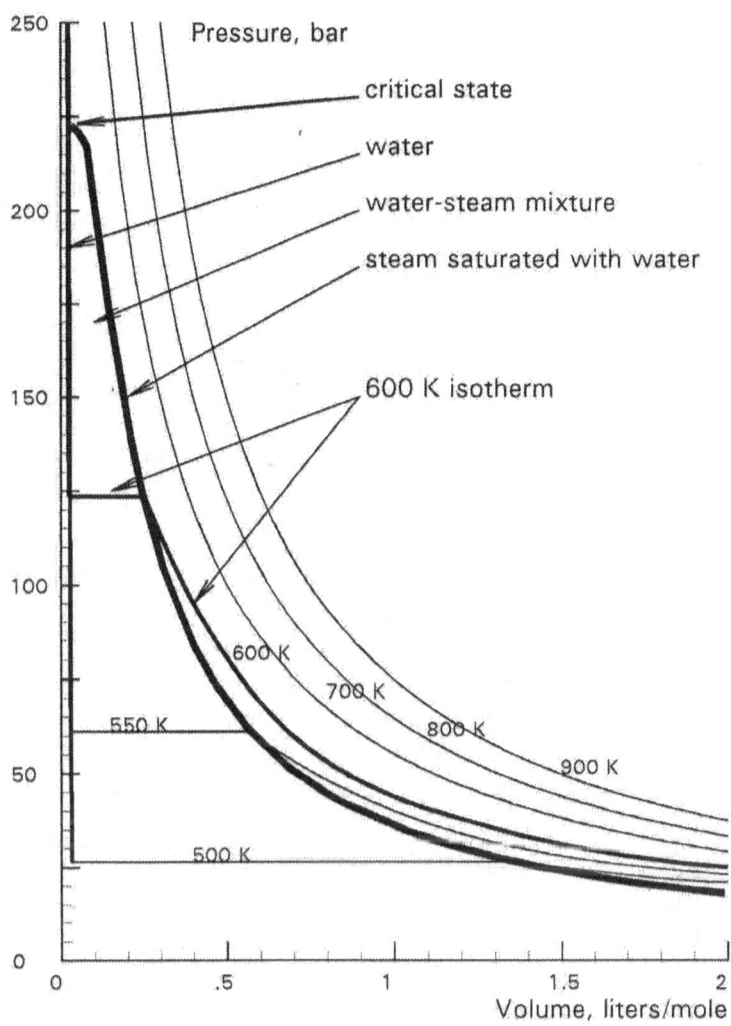

Fig. 7.5 Water-steam PVT phase diagram.

The Rankine Cycle Engine

Fig. 7.6 is a T-S diagram and corresponding PVT diagram of the exchange of heat and work done by the steps of the Rankine engine cycle. Like the gas turbine, the adiabatic expansions cover the pressure range faster than the temperature range leaving the output at a temperature too high to waste. Steam engines delay this inevitable loss by reheating the steam before expanding it to lower pressure by successive stages.

Table 7.4 summarizes the processes between the states numbered on the diagram. The Rankine engine cycle uses the PVT states shown in Fig. 7.4. The saturation curve in the T-S diagram shows that liquid water saturated with steam is pumped into the boiler and heated until it reaches the maximum pressure. The liquid then vaporizes until it is steam saturated with water. In the region above the critical point the distinction between the liquid and vapor disappears.

Table 7.4 Steps of a Rankine cycle engine.

1→2	Superheated steam expands adiabatically performing work by the 1st stage turbine.
2→3	Steam exhausted from the 1st stage turbine is reheated.
3→4	Superheated steam expands adiabatically performing work by the 2nd stage turbine.
4→5	Wet steam condenses at P(min) to liquid water at the heat sink temperature.
5→6	Liquid water is pumped into the boiler doing work that raises the pressure from P(min) to P(max).
6→7	Liquid water is heated until its vapor pressure reaches P(max).
7→8	The water-steam mixture is heated at P(max) until the water is completely vaporized.
8→1	Steam is superheated to the engine input temperature.

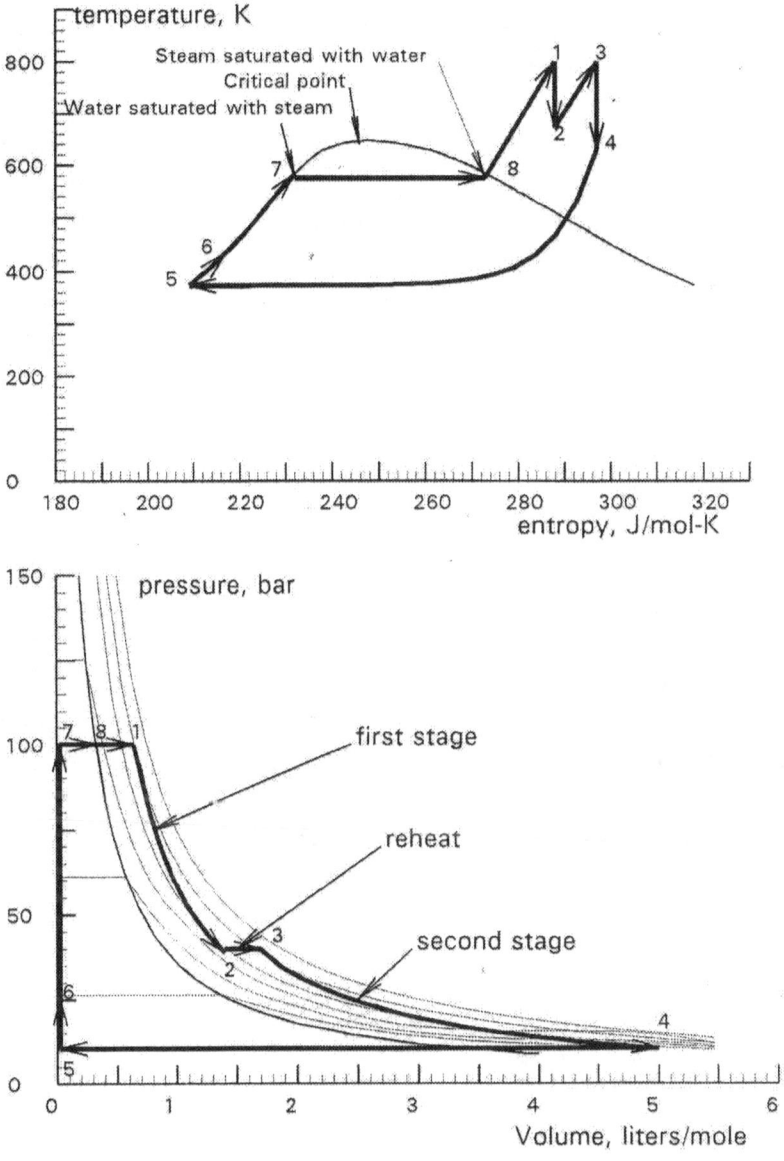

Fig. 7.6 2-stage Rankine cycle TS and PVT plots

The output of the pump that returns the condensed steam to the boiler at high pressure, state 6, is injected into a boiler where the temperature is higher. It is in fact much higher than the temperature at which the water can remain liquid. The temperature of the system in this state is therefore undefined, and the sequence of processes, 5→ 6→7, associated with it are irreversible. Although the sequence of states connecting these processes is undefined, the error in regarding states, 5 and 7, as equilibrium states is relatively minor.

The temperature and entropy limits of a Carnot cycle operating between the same temperatures was outlined in the T-S diagram of Fig. 7.2. The relative areas of the plots show that the ideal Carnot engine efficiency is slightly greater than 50%. The Rankine engine has about half that efficiency, giving the typical 25% efficiency in a single cycle.

Refrigerators, air conditioners, and heat pumps use a similar cycle with a different working fluid. Liquid refrigerant at high pressure is expanded isothermally, vaporizing the liquid, and extracting heat from the surroundings. The gas-liquid mixture is then compressed adiabatically to a high pressure. The heat the fluid absorbs in expanding, plus the adiabatic compression leaves it at a high temperature. The heat is exhausted to a heat sink at a higher temperature than the surroundings from which the heat was extracted, completing the cycle. The net result is that heat is extracted from one volume and pumped into another volume. The unit is a heat pump or a refrigerator depending on whether the focus is the volume that is warmed or the volume that is cooled.

Effect of Scale[7, 8]

This is the title of an essay by J.B.S. Haldane on limitations of structures natural selection creates. For example in falling from a cliff a mouse might be stunned, but otherwise unhurt. A human is broken, but the body remains intact. A horse splatters.

The scale of engineering developments must systematically recognize related considerations. There are always questions to answer before committing to build a full scale plant. It is necessary to consider how the principle changes with scale.

Geometric similarity describes systems in which all important variables are proportional to the geometric dimensions. The performance of power systems generally improve with scale. Small scale plants are not geometrically similar to large plants.

Dynamic similarity is the process used to design a small scale system that mimics a large scale system in some important respects. The design adjusts non-critical variables whose effect is not an issue to give the performance that is expected of the critical variables.

First consider a large scale steam boiler 10 m long by 4 m diameter insulated with 20 cm of insulation to reduce the heat loss. The heat in the boiler is proportional to the volume. The rate of heat loss is proportional to the surface area but inversely proportional to the thickness of the insulation.

Now consider a geometrically similar 1/100 scale boiler with the same dimensions in centimeters instead of meters. The small scale surface/volume ratio is the same as the larger boiler. The relative heat loss is 100 times greater because a geometrically similar insulation thickness is only 2 mm. A 1% heat loss in the large boiler becomes 100% in the small boiler.

Increasing the insulation thickness from 2 mm to 20 cm gives the two boilers dynamic similarity with respect to relative heat loss. However the volume of the insulation is now 100 times the volume of the boiler. The insulation contains more heat than the steam.

Another way to achieve dynamic similarity is to replace the insulation with a heater. By setting the heater at the 1% heat loss that is expected at the large scale the small scale apparatus is dynamically similar with respect to heat loss.

Nuclear fusion experiments are an extreme test of the principle of dynamic similarity in engineering research. In nature the Sun is the clear example of large scale nuclear fusion. It is the source of the energy from the Sun. It depends on the mass and gravitational field strength. Jupiter is not large enough to be part of a binary star.

Fusion occurs when adiabatic compression raises the temperature of a potential star to the million degree range that ignites fusion reactions. At sufficiently high temperature it forms atoms with atomic weight up to about 56 (iron). The fission products are slightly lighter than the component particles. The excess mass appears as kinetic energy of whatever particles are formed.

The principal fusion reactions in the sun are among various combinations of hydrogen, deuterium, tritium, and/or lithium nuclei to form a helium nucleus plus additional neutrons. About 4/5 of the 18 Mev of energy goes to the neutrons.

Fusion in a hydrogen bomb results when a nuclear fission bomb compresses the fusion reactants to their ignition temperature. The high energy of a hydrogen bomb is due primarily to the high efficiency resulting from greater neutron flux during the fraction of a second the nuclear reactions take place. Two different approaches are used in dynamic similarity experiments to determine the fundamental feasibility of fusion as an electric power source.

Inertial confinement fusion[9] is an experiment that mimics a hydrogen bomb on a micro scale. An explosive shell surrounds a spherical capsule containing a few milligrams of hydrogen-deuterium-tritium-lithium fuel mixture. A high energy laser pulse illuminates capsule uniformly from all directions. The momentum of the resulting implosion raises the temperature to a level that ignites fusion. The objective of these experiments is to understand the practical physics of an efficient ignition reaction. It does not address the issues concerned with confining, capturing, and sustaining the reaction.

Magnetic confinement fusion[10] is an experiment that first converts the fuel mixture to ionized plasma. On injecting the plasma into the intense magnetic field in the cavity of a *Tokomak magnet* the system becomes a surrogate for the gravitational field of a star. The system that generates continuous power depends on the outcome of further developments.

Each of these approaches has advanced through a sequence of multibillion dollar improvements over several decades. Both approaches have made a series of incremental improvements in the efficiency of the ignition reaction. The magnetic confinement experiments have passed the break-even point where the fusion energy exceeds the energy needed to create the field. They have provided information that shows a path to 50% efficiency and higher. Similar incremental improvements to extend the duration of the reaction from momentary to continuous operation are in progress.

Stars are heavy enough for gravitational forces to ignite fusion reactions and sustain the ignition temperature. They are also large enough to capture most of the energy of the fusion products and sustain the reaction over a long stellar lifetime.

In a nuclear fission power reactor the temperature of the reactor is modest and high temperature is not required to sustain the reaction. By the same token the heat of the reaction products is easier to capture in surroundings that moderate a critical mass of modest size. The nature of a critical mass that can moderate a reaction at a 10,000,000 K and transfer heat to an engine at 1000 K is yet to be determined.

Notes and References for Chapter 7

1. The experimental basis of thermodynamics was developed for the most part by European scientists. Josiah Willard Gibbs, Professor of Mathematical Physics at Yale University and perhaps the only outstanding American theoretical scientist before the 20th century, established the theoretical basis. His work was first published as Transactions of the Connecticut Academy of Sciences in 1875-1876.

2. Gibbs' work clarified chemical equilibrium by introducing the concept of chemical work, $\mu_i \, dn_i$, where μ_i is the *chemical potential* and dn_i the change in the number of moles of substance i in a chemical mixture. During the 20th century chemical potentials (partial molar free energies) for enough substances have been determined to make equilibrium composition calculations routine.

   Table of Thermodynamic Properties, National Institute of Standards and Technology

3. The Pascal, Pa, $(kg\text{-}m/sec^2\text{-}m^2)$ is 1.01332×10^{-5} times the standard atmosphere.

   One bar, 10^5 Pa, is about 1% greater than the standard atmosphere.

4. Note that some standard states are unattainable. Water does not exist at 298 K and 1 atmosphere although that is its standard state.

5. Saturation curve data, tabulated as Steam Tables, are given in power engineering handbooks. L. Haar, L., and J.S. Gallagher, and G.S. Kell, *NBS/NCR Steam Tables*, Hemisphere Press, New York, 1984

6. The critical constants for water are $T_c = 647.14$ K, $P_c = 220.6$ bar, $V_c = .056$ liters/mol. At temperatures and pressures above the critical point the distinction between liquid and gas vanishes and the system is called a *fluid*. As substances approach their critical state the meniscus which normally separates a liquid and gas disappears. At the critical state the entire system suddenly becomes opalescent. This phenomenon called *critical opalescence*.

7. *On Being the Right Size*, a collection of essays by J.B.S. Haldane (1892-1964), has been edited and extended by John Maynard Smith, Oxford University Press, 1985

8. *On Growth and Form* is a classic work by D'Arcy Wentworth Thompson (1860-1948) describing natural "engineering tests" of the relation between size, growth, and function of plants and animals that survive natural selection. It was abridged and reprinted by J.T. Bonner, Cambridge University Press, 1961.

9. The National Ignition Facility at the Lawrence Livermore National Laboratory is the culmination of three decades of inertial confinement research support by the U.S. government.

10. Magnetic confinement research by universities and national laboratories of several nations will now culminate in joint effort funded by China, E.U., India, Japan, South Korea, Russia, and the U.S. known as the *International Thermonuclear Energy Research* (ITER) project. Construction is scheduled to start at Cadarache France in 2008.

Chapter 8
Requirements of a Power Grid

The electric power industry began in the United States as privately owned businesses that generate power and distribute it to local consumers. Electricity flows wherever there are connections. The demand for power spreads by adding companies that distribute power to new areas, companies that generate power, and companies that transmit power.

A vertically integrated monopoly is a natural economic model for an electric power system. Consumers are hard wired to the distribution grid. The distributors who build, maintain, and own the grid require many years of revenues to amortize the capital cost. Distributors require reliable power in the required quantities at the lowest possible cost. This requires equally long range planning by plants which achieve low cost by a large scale operation. Governments tend to accept the inevitable monopolistic necessity and protect the public interest by regulating prices to consumers[1].

An alternative economic model in which a large number of small power sources contribute to the same grid might apply to renewable energy sources such as wind and solar power. These sources, which might be distributed as broadly as individual residences, would fit a more market oriented system that might supersede the nonrenewable sources.

The *U.S. Comprehensive Electricity Competition Act of 1999* separated power generation and distribution businesses in the U.S.[2]. How this affects the economics of electric power and whether it will help develop a market for energy alternatives is likely to remain in flux at least until alternatives can deliver enough power to have a significant impact[3].

The electric power market has physical existence only where the grid connects to a potential source of immediate power. Market economics and other regulations are meaningless unless they satisfy the physical conditions that allow power to flow when it is sold. This chapter is a brief survey of those realities.

Table 8.1 reconciles the average output by power generators with the consumption by power users in the U.S. during 2002. The left hand column summarizes the energy provided from different sources. The right hand column summarizes the final disposition of that energy.

Table 8.1 Rate of energy consumption for electricity in GW[4]

Power sources		Power consumers	
Coal	585	Residential	136
Natural gas	103	Commercial	117
Petroleum	26	Industrial	122
Nuclear	240	Other	13
Hydroelectric	96	Sub-Total	388
Other	11		
Non Utility	67	**Lost power**	
Total	1128	Waste heat	690
		Plant Use	18
		Transmission	29
		Sub-total	737
		Total	1125[5]

About 85% of the total comes from sources that produce heat, convert the heat to mechanical power, then electric power. They are the nonrenewable fuels, coal, gas, oil, and uranium. Typically thermal energy sources can convert only a third of the heat to mechanical energy. A part of the two thirds of the energy that is unused might be useful for lower temperature heat. This is practical only for short distances.

The major non-thermal energy source is hydroelectric power. It is a renewable natural energy resource that is available in limited locations. The total hydroelectric capacity is large in comparison with the alternative renewable sources that now contribute about 2% of the total.

Less than a dozen alternative energy sources have potential to supply the grid. Each source has many attributes that determine how or whether the grid can use them. The nonrenewable sources supply most of the power primarily because they meet the conditions required by the grid, not because they have a monopoly. Many features are obstacles to a smooth transition.

The attributes of the four nonrenewable sources form the design specifications of the grid. This omits consideration of their uncertain future.

- They deliver power in sufficient quantity to meet the demand.
- They can each be located within transmission line distance of almost any location.
- They can each supply power at any time of day or season of the year on sufficient advance notice.
- Their flexibility to change output power varies with the technology. Their inertia increases in the order gas, oil, coal, and nuclear power.
- They can reliably meet contractual commitments made months in advance to deliver power within a stipulated range.

Renewable power has not been slow to develop by accident, neglect, or monopoly power. All of the renewable sources except space solar power are deficient in most of the attributes that make the nonrenewable sources major contributors.

The environmental hazard and the potential remediation costs are generally greater for nonrenewable than renewable energy. They increase in the order gas, oil, coal, and nuclear. Their cost of fuel handling results in higher cost of operation and maintenance.

The capital cost of a facility that delivers electric power to the grid on a meaningful scale is generally a major part of the total cost of power. Experience is a reliable guide to the cost of new plants using familiar technology. The capital cost of emerging technologies depends on the ingenuity of the designer and the development that follows. The discussions of alternative technologies in Chapter 9 present capital costs as targets that are allowed by revenues and operating assumptions.

Effect of Time-of-Demand on Price

Fig. 8.1 shows average daily distribution of demand per gigawatt of maximum capacity. Proven ability to provide quantities of reliable low cost power that are comparable to the peak demand distinguish coal, nuclear, and hydroelectric power as base power providers. They use economies of scale to deliver power at a low price. Although they can deliver power at any time they have limited flexibility to change output. Coal and nuclear plants require several days to go into operation from a cold start.

The base power indicated in the drawing includes the minimum standby power base power generators must produce to remain in operation. The maximum power a generator can deliver is its *nameplate power*. This is the top of the shaded area of the curve. The shaded area between the nameplate and the base power is *spinning reserve.*

Spinning reserve is the range in which simple adjustments of input to a large base power generator are possible. The entire area of the shaded portion of the curve is the power that base power generators must be able to provide.

The utilization ratio of a particular power generator is the ratio of the average power to the nameplate power. It is 74% in this example. The range above the shaded area is *peaking power* that is supplied by a more flexible generator. Even if base power generators have enough spinning reserve to supply all of the power doing so would require low utilization of an expensive resource.

The peaking power generators in wide use for this purpose are gas turbine engines. They operate on natural gas fuel. They can be brought on line from a cold start on about 15-minutes notice. In this example peaking power generators furnish 12.7% of the power with 28% utilization factor.

The capital cost of a plant is a major fraction of the cost of power. Plant utilization is thus a major factor in the price. This example illustrates how a combination of low cost inflexible sources and high cost flexible sources can produce reliable, low cost power on demand.

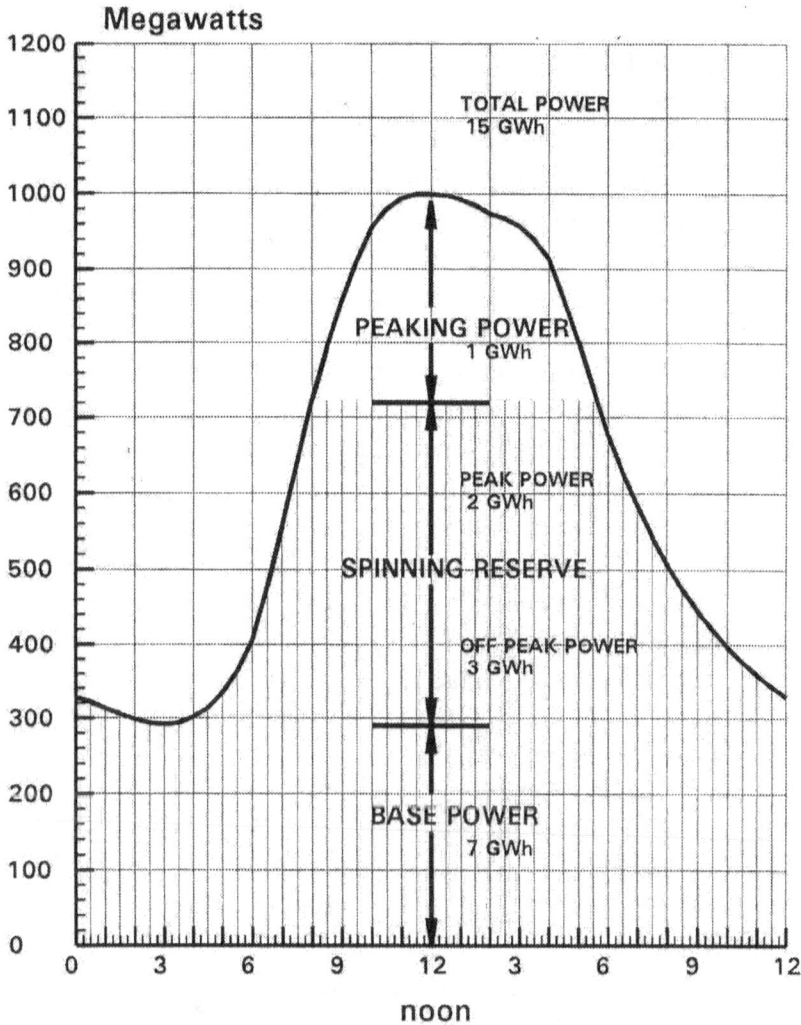

Fig. 8.1 Classes of daily power demand

Fig. 8.2 shows the seasonal distribution of electric power. Most of the differences can be attributed to different classes of consumers.

Residential consumer demand has the greatest variation. The high summer demand is attributable to air conditioning. The smaller peak in January is attributed to the higher consumption for heat circulation and longer nighttime illumination.

Commercial demand mirrors residential demand with about half the magnitude of variation. Large consumers who use insulation and other measures to control the timing of their demand negotiate better prices.

Industrial power demand is generally less variable than residential or commercial demand. It is often a factor of production. Industries like electrolytic aluminum can well afford to adjust production schedules to minimize cost.

Utilities consume electricity for public purposes such as water supply, sewerage disposal, and streetlights. Late night lighting may qualify for special base power rates.

Time-of-use pricing introduces a market distinction in the relation between cost and value of power consumed during periods of maximum or minimum demand. The peak generating capacity is determined by the peak demand at mid-day during July. This is more than 4 times the late night demand during November.

Real time metering can adjust prices at intervals less than an hour. Large consumers negotiate special rates by adjusting their consumption to quantities and timing that are predictable. This gives producers a way to maximize utilization of their capital facilities.

The minimum price is for power that can be anticipated a day or more in advance. The power required by unanticipated real time adjustments can have prices that are higher by a multiple of the basic rates.

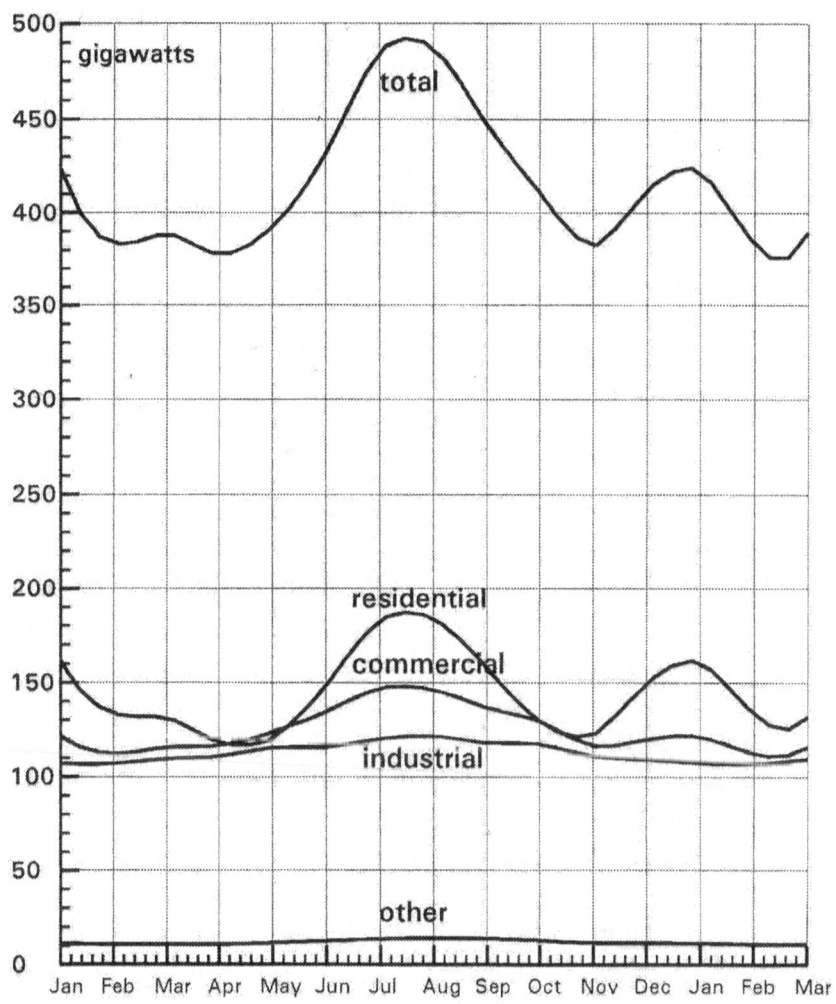

Fig. 8.2 U.S seasonal electric power demand

Organization of the Grid

Fig. 8.3 is a schematic illustration of the connections to individual consumers that a local power distribution company might own and maintain. It includes connections to receive power from one local area power plant from a regional distribution company. Describing the grid as a network in which every part is ultimately connected to every other part can be misleading. The grid is more realistically a hierarchy of connections and responsibilities that depends on the reliability of each part[6].

Local service operators buy power from a limited number of sources for sale to individual consumers. They are responsible for the part of the grid that connects power to a specific set of consumers. This includes maintaining the flow of power as well as the physical connections to the consumers. The sources may include a direct connection to one or more local power plants.

Larger regional power distribution companies more often have direct connections to several power plants. These connections can be either direct or through lines of high voltage transmission companies. Their customers may include local service operators as well as their own grid of local consumers. The grid is essentially passive except for input switches to suppliers and output switches to consumers. The consumers have primary control of the output switches. The power distribution companies have control of the input switches.

The current carrying capacity of wires is small in relation to the power they must deliver. This is the problem Tesla resolved with alternating current. The grid distributes 3-phase ac power at a range of voltages consistent with the applications of different classes of consumers[7, 8]. The heaviest lines represent very high voltage transmission lines. Progressively lighter lines deliver smaller quantities of power at lower voltages. The 110 volt single phase power used by individual residences is usually provided by a transformer at or near the site.

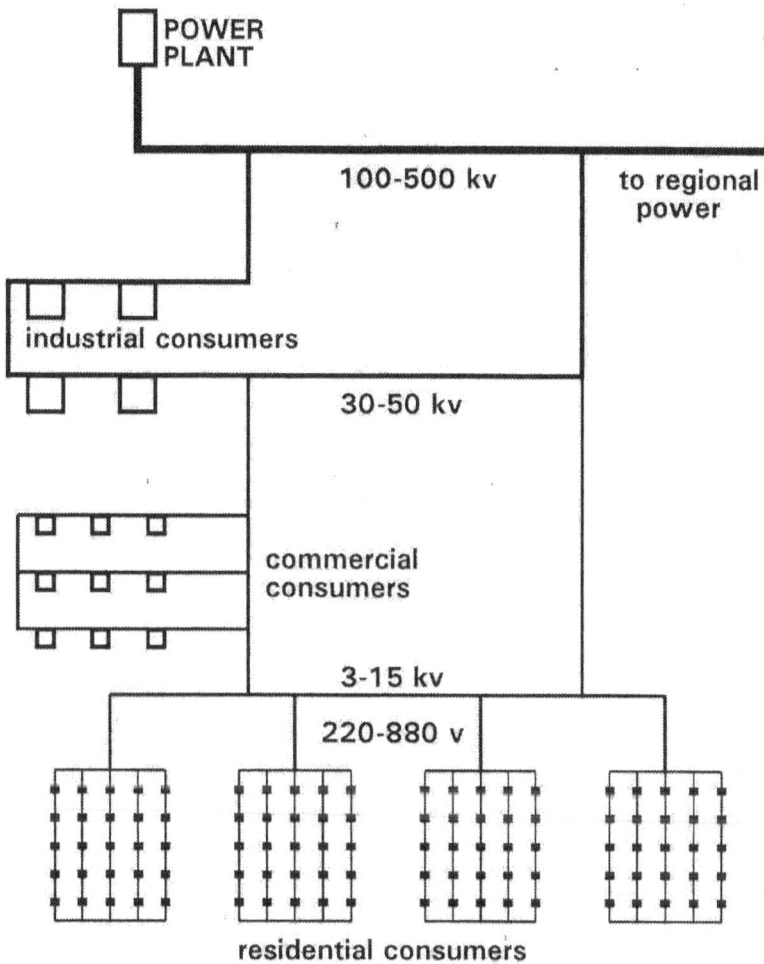

Fig. 8.3 Schematic power grid connections

Operation of the Grid

A power grid operation has some similarity to an airline traffic system. Power consumers produce an aggregate demand close to that which is predicted a day or more in advance. The demand follows a rough schedule predicted in advance by the season of the year, time of day, and other anticipated events. It has shorter range predictability by contemporary events such as weather. In addition it has a real time component that is unpredictable.

Power plant operators are the pilots in command. After reaching cruise conditions they monitor the power production, make routine adjustments to maintain the scheduled output, and make whatever other responses that are appropriate for changes in demand outside the range of the routine schedule of adjustments.

An automatic governor on each power generator corrects small deviations from nominal by a throttle controlling the steam or other heat flow to the engine. The margin of safety between the nominal demand and the limits of the spinning reserve for that particular power plant operation is the critical information to monitor. It determines what response the power generator can make to any unexpected deviation from the nominal schedule.

Distribution center operators monitor the flow of power in all parts of its component of the grid. They have little physical control of operations other than the drastic measure of disconnecting specific consumers or power generators from the grid. They issue clearances to change the nominal output dispatch with the consultation and approval of the power sources. The power source must confirm that the change will not conflict with good operating procedures and standards. The clearances have the effect of authorizing a tentative sale of power.

Regional distributors function with more numerous power sources and fewer end users. Regional transmission organizations are non-profit companies formed as a power pool to assure adequate wholesale power for an entire region. The law limits what fraction of the generating capacity a single company can control.

Merit order dispatching according to the lowest current price offered by a legitimate owner of a capacity to generate power has priority for meeting increased demand in the absence of other priorities. Historically the Federal Energy Regulatory Commission set U.S. rates by outlining broad principles and by approving specific kinds of *power purchase agreements*. These rates allow generators, transmitters, and distributors to recover cost plus a reasonable return on the capital investment[9].

The day-ahead sales are generally bilateral contracts and sales agreements that limit financial transactions to those actually responsible for the physical transactions. Selling power in advance implies a firm ability to deliver it at the time of demand. Control of the grid and the corresponding electric power sales are divided according to the day-ahead demand which can be planned long in advance and real-time demand which may require hour-ahead adjustments. This does not necessarily produce the decade-ahead stability required for low lost capital financing.[10]

The theoretical *efficient market* requires (a) equal transparent access to all information relevant to the cost and value, and (b) equal access to a wide choice of buyers and sellers. If the number of transactions is large enough to be statistically significant the cost and value then converge to define the fair market price.

The Open Access to Same-Time Information System (OASIS) is the U.S. effort to form a transparent market. It reports all sales of electric power, the time of the sale, and the identity of the seller and buyer. Unlike other commodities, electric power exists only at the moment it is generated. Marketers can claim to own future power they never control. Even if the power exists its ownership must be backed by a history of physical control of when, where, and how much power is delivered.

Current congestion is a major concern of distribution center operators during periods of high demand. The grid itself is essentially passive. The magnitude and even the direction of the current depends only on the voltage difference between the power source and the consumer load. The first step is to be aware of conditions that would cause congestion to escalate.

The main tool available to a controller is information about the state of power flow throughout the grid in relation to the capacity of the transmission lines to carry the current. The load consists of two kinds of current. They have equal weight in taxing the capacity of the system but have quite different effects in contributing to the power load.

Real power is consumed by resistive loads that produce heat or useful work. It is the power recorded by the usual current meter. Fuel energy corresponds to real power that is the essential element of the cost of electricity.

Reactive power is current that charges the inductance or capacitance of reactive loads. This is typically the inductance of large electric motors or ballasts for fluorescent lights. The current also has a real component that produces work, radiation, or heat. It is power that is borrowed for half a cycle at a time and not consumed. It is distinguishable from real power and is a metered element of the price to large consumers.

Inductance or capacitance shift the phase of the voltage with respect to current. This stores, then releases power at each reversal of the cycle. Reactive power is expressed in VARs (volt-amperes, reactive). It appears as a deviation from the nominal phase. Large consumers eliminate reactive power by adding capacitance to shift the phase in the opposite direction. Full compensation gives a resonant circuit that draws only resistive power.

Wheeling power passes through the lines of an independent system operator without being bought or sold. P*ass-through transmission* nevertheless has costs that can fairly be charged to the buyer, particularly if it results in congestion that requires the local operator to deviate from the planned delivery schedule.

Shedding load may be the only option a distribution center operator has left when congestion approaches the safe limits of the grid. Each distributor has a list of major power consumers with the priority of their service. Power sales to these consumers are negotiated. Time-of-use and priority-of-service are elements of the rates that are included in the contract.

Circuit breakers are distributed throughout the grid to protect all major equipment from accidental power overloads. Many of them are equipped to close and restore power automatically if the fault is temporary. A large fault produces a power glitch in the form of a damped oscillation at the resonant frequency of that section of the grid. A temporary glitch, such as that which occurs on adding or shedding a large load element, typically damps out in a dozen cycles or so. If not, the voltage step can propagate throughout the grid.

A margin of safety large enough to absorb any failure that might reasonably occur is the only real protection for reliable power. A fault occurs far too quickly for an operator to have a meaningful response, let alone analyze what the response should be in relation to where the fault occurred. Balancing the margin of safety throughout the grid is one more element to consider when dispatching additional power to cover an increasing demand.

Thyristor switches make it possible, in principle, to design a "smart grid" that automatically optimizes the grid performance in real time. This would require objective, unambiguous criteria for ideal grid performance[11].

Thyristors can provide feedback stabilization to control the voltage, frequency, and phase of a generator. They can convert ac voltages to dc and vice versa to permit high voltage dc current transmission. They can change transformer output taps to change live transmission line voltages in real time. They can introduce active phase compensation at any point in the grid. They can open a circuit with an uncorrectable overload, such as a short circuit, and automatically restore the circuit if the overload turns out to be temporary.

Complex methods of fast electronic control are an emerging technology. Controlling reactive power transients in transmission lines requires more extensive real field experience to meet high reliability standards with no unintended consequences[12].

Alternative Power Compatibility

The grid now has a small number of very large sources. One model for the growth of alternative sources is a grid with a vast number of widely distributed small sources of wind, solar, and other forms of cogeneration for both private and public use. By the air traffic analogy this would be like replacing most of the present fleet of airliners with private aircraft. It is not clear how or whether this is feasible. It is unlikely to occur suddenly. It is not too soon to consider the physical requirements.

The insignificant contribution by alternative energy sources has less to do with economics than their failure to meet the requirements of the grid. The present use of alternative power is possible mainly because their fraction of the total is small enough for the grid to tolerate the problems they introduce. The number of alternatives is limited. Chapter 9 examines them individually in detail. The general question is whether they have the aggregate magnitude to solve the supply problem. If so, what adaptive changes by the both the grid and the sources themselves are necessary.

The aggregate statistical average supply must deliver power with the predictability a contract for firm delivery requires. The absolute physical requirement is for the supply and demand to coincide. This would presumably require something equivalent to computer control of the spinning reserve for each source.

A rigorous physical requirement of the grid is that electric power generation and consumption occur simultaneously. Failure to contribute power during periods of peak demand is the most serious weakness. The California heat wave in the summer of 2006 illustrates the conflict.

A wind turbine can produce up to 1000 kW of power at nearly competitive cost. California encourages renewable power with guarantees to buy it. About 2.5 GW of its 50 GW total capacity is now wind power. Although wind generally rises during the day it often dies down during a heat wave. Power demand for air conditioning goes up.

During the heat wave of 2006 the demand rose to the peak capacity. The contribution from wind fell first to 0.7 then 0.1 GW. Failure to contribute to the peak demand is only part of the problem. At other times wind competes with other power sources, reducing their utilization.

The problem is not unique to wind. Solar power generally agrees well with the daytime demand, but fails at night or whenever the sky is cloudy. Solar power with its peak at noon is a good match for the peak and near peak demand. To avoid interruptions by weather it needs an arid location. To minimize seasonal variations it needs low latitudes. Wind power varies more widely geographically as well as in timing. Neither solar nor wind power can guarantee a firm contractual output. In a statistical combination they would do better.

Time-of-use pricing is a way for the market to influence demand by a price based on demand. Each hour of the year falls into a demand class by time of day or season of the year. The far off-peak demand covers 10% of the hours shortly after midnight. Producers now sell this power below cost to keep the *must-run* generators operating. Off peak demand is low value power during 30% late night hours. The near-peak demand covers 50% of the hours of daylight and early evening. The peak demand covers the 10% of the hours near noon. Base power sources have the capacity to supply all of the power below the peak demand. Sources with more flexible response supply peak demand at a higher price.

The value of a power source depends on whether it can supply a meaningful quantity at the time of the demand. Every source must be part of a plan with sound economics. In addition it must answer the following questions.

- What fraction of the total demand can it supply?
- Can it contribute to the peak demand?
- Is the contribution reliable?

A statistical combination of many small sources with a broad geographical and technical diversity might have statistical reliability. An historical record over time would give the data needed to plan it. For the statistics to help, a broader variety of cogeneration sources needs to come into play.

Inter-regional transmission can *shift load* from a region with adequate power to a region with inadequate or higher cost power. Although a high voltage transmission line cannot create power, it has characteristics that might support the statistical diversity of distributed power sources at a cost.

Transmission lines do not create power. The capital cost depends on utilization much like the capital cost of generation. This adds to the cost to the consumer. Transmission lines are least reliable during periods of peak demand and require a significant margin of safety. They are passive systems with limited control during an emergency.

Transmission line losses increase with distance but are tolerable below about 500 to 1000 km. This might give consumers at a particular location a large fraction of a million square kilometers over which the diversity of power sources might be averaged for reliable delivery. This would require a much higher order of grid management technology than is now available.

Notes and references for chapter 8

1. In the U.S. perceived abuses of monopoly powers led the *Federal Power Act of 1935* to establish the *Federal Energy Regulatory Commission* to regulate electric power sales. Its jurisdiction is constitutionally limited to interstate commerce with local regulation left to the states. However increasing interstate power traffic increases the federal jurisdiction accordingly. Power was regulated primarily by setting rates for different classes of consumers that satisfy a revenue requirement which includes operating cost, depreciation, taxes, and a reasonable return on investment.

2. United Kingdom, Norway, and Germany have all accomplished various kinds of privatization and deregulation. However, since the markets in these countries are smaller than most U.S. regional power pools, they are not a close analog of the U.S. market. William L. Hogan, *Making Markets in Electric Power*, notes for a Cantor Lecture before the Royal Institute for the Encouragement of Arts, Manufacturing, and Commerce, London, 2000. Reprinted as Chapter 4 in *Governance Amid Bigger and Better Markets,* John D. Donahue, and Joseph F Nye, Jr,, Brookings Institution Press, Washington D.C., 2001

3. In separating power generating and distribution the *Comprehensive Electricity Competition Act of 1999* eliminated the vertically integrated monopoly but the distribution companies remain a monopoly to be regulated. James H. McGrew, *Federal Energy Regulatory Commission*, American Bar Association, 2002

4. Monthly Energy Review, Energy Information Administration, U.S. Department of Energy, May 2003

5. The mismatch in totals is from components of power generation not reported as consumption of fuel for electric power. The data are a summary for 2001. *Annual Energy Review 2003*, Energy Information Agency, U.S. Department of Energy

6. John Casazza, *Understanding Electric power Systems: An Overview of the Technology and the Marketplace*, Wiley-Interscience, 2003

7. Transformers exchange voltage for current at a particular power. The input current passes through a *primary coil* that produces an alternating field in a magnetic core. The field strength is proportional to the current and number of turns. This induces a current proportional to the number of turns in a *secondary coil* around the same field. The primary/secondary current ratio is inversely proportional to the turn ratio. The voltage ratio is directly proportional to the turn ratio. The power loss which appears as heat is generally small. High voltage lines typically transmit power by three conductors that carry current at phase angles that differ by 120°. A fourth ground reference carries no net power.

    Three-phase current provides smooth power to loads, such as heavy induction motors. Aluminum has largely replaced copper as the electrical conductor. Cables of super-conductor material that approach zero theoretical resistance at cryogenic temperatures are an emerging technology that is now practical for power transmission over short distances. Fabio Saccomanno, *IEE Electric Power Systems, Analysis and Control*, Wiley-Interscience, 2003

8. The original transmission line between Hoover Dam and Los Angeles was 6 parallel 1.4-inch diameter cables with central steel strands for strength surrounded by hollow copper strands to carry current. Each cable carried 500 amperes at 287,000 volts in two 3-phase 60-cycle circuits separated by 32.5 ft. The line delivered 860 MW at over the 266-mile distance with 5% loss.

9. The deregulation process is set forth in Rules 888 and 889 of the Federal Energy Regulatory Commission. In its review of these rules the U.S. Supreme Court said, "The Nation's electric supply epitomizes interstate commerce and cannot be effectively regulated by individual states." At a given current a dc transmission line can operate at 40% higher voltage than an ac line.

10. Gregory V.,Welch, Michael V.Engel, , and Harold W.Adams, Jr., *Acquiring Energy Resources in a Competitive Market*, IEEE Power and Energy, **1**, 36, 2003

11. R. Mohan Mathur and Rajiv K. Varma, *Thyristor Based FACTS Controllers for Electrical Transmission Systems*, John Wiley & Sons, 2002

12. Melvin Olkin, *The Grid in Transition*, IEEE Power & Energy, **1**-5, 88, 2003

Chapter 9
Renewable Alternatives

Separating power production from distribution gives renewable energy sources equal access to the power market. Whether this matters depends on the properties of each individual source. The primary question is whether a statistical combination of renewable power sources can satisfy requirements that enable the grid to meet the demand.

Table 9.1 summarizes the Earth's source of renewable power[1]. Radiation from the sun accounts for about 99.8% of the total energy. Radiation input to the cross section exposed to the sun is balanced by thermal radiation from the entire surface 24-hours a day[2].

Table 9.1 Earth power resources (in petawatts, 10^{15} W)

Solar	173.
Earth core	0.2
Tides	0.001
Total input	**173.2 PW**
Reflected albedo	52.
Thermal radiation	121.
Surface heating	80.
Evaporation - precipitation	37.
Air circulation	3.0
Photosynthesis	0.18
Total output	**173.2**
World consumption (2000)	0.017

Solar radiation provides the overwhelming majority of all the energy flux that is available at the Earth surface. This chapter examines processes such as evaporation-precipitation and bio-fuels that provide temporary storage of solar energy for short periods as well as energy sources other than direct sunlight. Chapter 10 considers the methods of using solar power directly.

Present Use of Renewable Power

The present operation of the power grid is based on two kinds of producers. Bulk producers provide the base power from plants designed for large economies of scale and low price. But they are inflexible to changes in demand. Peaking power producers give fast, reliable response to changes in demand at significantly higher prices. Although more sustainable alternatives will ultimately be necessary they have been slow to develop.

Fig. 9.1 compares present production by renewable power sources. They combine to provide about 10% of the total electric power. Hydroelectric power is the only significant non-thermal source that is a base power provider. At present it provides about 75% of the renewable power. In the U.S. its share has fallen from its peak of over 20% to about 7.5% because the most promising sites are already developed.

Renewable sources other than hydroelectric are each relatively insignificant by comparison. Power from biomass is mostly gas produced by the tertiary treatment of sewerage which is now mandatory. Power from wood is significant mainly because the waste products of forest industries are too valuable to discard. The 0.5 GW contributed by direct solar power is omitted because it is less than the thickness of the line on the graph. Much of what little growth these technologies have experienced has been because their contribution is too small for their shortcomings in filling the requirements of the grid to matter.

Distributed power from a large number of widely distributed small sources is a third possible component of the power grid organization. There are two major doubts about them. It is not clear whether they can provide enough power to meet the demand at a reasonable price. It is not clear whether any statistical combination of them can make up for their inherent unreliability. The first step in resolving these doubts is to examine the individual characteristics of each technology.

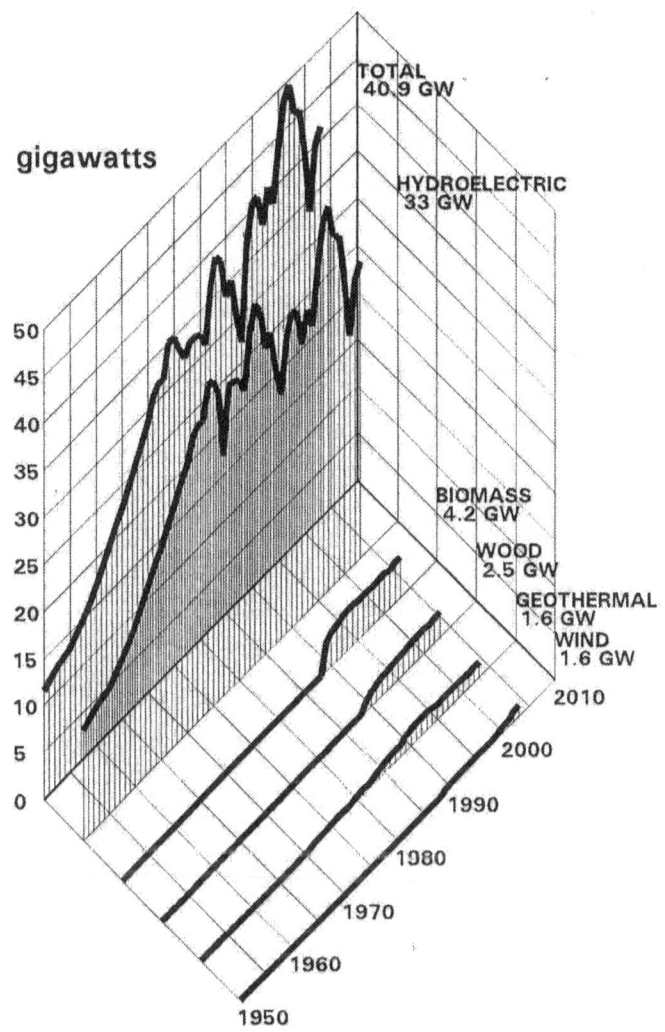

Fig. 9.1 U.S. renewable power consumption

Data Source: U.S. Energy Information Administration, AER 2004

Hydroelectric Power

The energy for hydroelectric power is the potential energy of liquid water due to relative elevation or *head* at which it was precipitated. To generate electric power a turbine captures the flow kinetic energy the potential energy produces to drive a generator. The energy is modest. One kilowatt-hour of electric power requires 400 metric tons of water divided by the head that creates the flow[3]. This ranges from ten thousand to a million times the mass of coal per KWh.

Table 9.2 shows that a hydroelectric plant with the same base power revenues as coal can justify a capital cost of almost $4.5 billion. Both the absence of fuel cost and simpler operation reduce the operating cost. Extracting power from a source with a low energy density requires large engineering structures at a relatively high capital cost. These structures can be amortized over longer time except for replacements of generating equipment.

Table 9.2 Cost of hydroelectric power

Revenues	
Nameplate power	1 GW
Utilization	.60
Price	$0.06 / kWh_e
Annual electric energy	5256 GWh/yr
Annual revenue	$315 million
Operating cost	
Total operating cost	$50 million
Capital cost	
Annual payment on capital	$265 million
Rate of return on investment	0.08
Amortization period	50 y
Allowable capital cost	$4422 million

Three main types of power plant all use some aspect of the Bernoulli principle to concentrate the flow energy. Water turbines typically convert the potential energy of the massive flow to electrical energy with about 85% efficiency.

A dam and reservoir power plant uses a deep narrow canyon that can form a deep reservoir with a large storage volume and head. The height of the dam determines the head of water at the base to be converted to electrical power.

A water diversion power plant, like Niagara Falls, diverts water at the upper elevation through a tunnel to a lower elevation. This produces just as much head, but relies on the flow continuity in lieu of storage. The diversion may shift water to a different watershed for public water supply or irrigation.

A run-of-the-river power plant uses a dam as a weir to increase the normal linear flow velocity of the river by decreasing the flow cross section. Like a diversion plant, this relies on the continuity of the river flow in lieu of storage.

Kaplan reaction turbines resemble a ship's propeller. They operate at a low axial flow rate using a small blade angle. They are useful at sites with a low to medium pressure head. At higher flow rates the blade angle is increased to allow water to contact the blades through a longer distance.

Pelton impulse turbines capture the energy from higher dams in radial buckets. The blades are shaped to convert the linear momentum of the water to angular momentum of the turbine.

Most hydroelectric power systems are multiple use projects for civil water supply and crop irrigation as well as power. The Grand Coulee Dam power is used for pumped storage that fills the Grand Coulee, a dry canyon above the river. This irrigates about a million acres in western Washington. Similar projects on the Colorado River allow large areas of relatively uninhabitable land to support large and productive populations.

The rain collection and flow volume of the major river systems is at the relatively flat lower elevations. The hydroelectric power that is economical to develop is usually a small fraction of the watershed in the upper part of the river's origin. This resource is the top 300 GW out of perhaps a total 3000 GW. Europe, North America, and Australia have developed most of the sites that are economical to develop.

Table 9.3 summarizes hydroelectric power generation in selected nations. The number in parentheses is an estimate of the total capacity when projects planned or in progress are complete[4].

Table 9.3 Hydroelectric power consumption in selected nations

	$W/person	% hydro	GW hydro	
Argentina	275	32	3.3	
Australia	1250	8	1.9	
Brazil	832	87	34.8	(54)
Canada	2229	59	40.9	(67)
China	122	16	25.4	(65)
Finland	1598	21	1.7	
France	1037	13	7.6	
Germany	789	4	2.5	
India	60	14	8.5	(25)
Italy	531	16	5.1	
Japan	972	8	10.0	
Mexico	236	16	3.8	(5)
Norway	3251	100	16.2	
Russian Fed.	690	19	18.7	(21)
Sweden	1851	54	9.0	
United States	1604	6	28.3	(34)
World	286	17	304.8	

Source: World Bank, World Development Indicators 2002.

The U.S. Grand Coulee Dam on the Columbia River, with a capacity of 4.2 GW, the world's largest when it was built[5], was part of a series of public developments of the Columbia, Colorado, and Tennessee River rivers which were justified as multiple use power, water, and economic developments.

The Three Gorges Dam on the Yangtze River in China is designed to produce 18.2 gigawatts from 26 generators with a 175 m head[6]. The 630 km² reservoir allows 116,000 t/s of flow. China is also developing 40 GW on the Yellow River. India has enormous hydroelectric potential on rivers of the Himalayan watershed. Its development plans are even more ambitious than China's as a percentage of its total power, but less than half on an absolute scale. Potential sites totaling about 84 GW have been identified.

Projects that produce significant power not only convert desert land to crop land but change the ecology of the river system. Political and environmental disruption follows from the large amount of land submerged or altered.

Hydroelectric turbo-generators are efficient. They are also reversible. They can capture water flowing from a reservoir to generate power. By reversing the electrical current they can pump water to a higher reservoir.

Pumped storage facilities pump water to a higher elevation reservoir. This stores low value power during periods of low demand and regenerates at the time of peak demand when it has higher value. Pumped storage might also store power from undependable sources to insure dependable delivery. The cost of pumped storage is a premium 130% of the base cost of the power. This is the cost of the storage facility utilization plus 85% efficiency in each direction. A large head is desirable to limit the quantity of water that must be stored as well as for efficient pumping. This requires a site higher than the outflow, not necessarily a deep reservoir.

The Lundington facility on Lake Michigan is one of the largest pure pumped storage facilities. It has a 1.9 GW capacity and can store up to 15 GWh of energy in a 4.8 km^2 reservoir with a 105 meter head. Unlike hydroelectric power, pure pumped storage does not change the ecology of an entire river system.

Pumped storage has a longer history of use in Europe than in North America. Governments favor developments that maximize the value of existing power resources[7].

Electric Power from Wind

Wind power originates as the inertia of the atmosphere with respect to the rotation of the Earth. The Coriolis Effect curves the horizontal component and daytime solar heating adds local vertical components to produce the average somewhat chaotic pattern that is observed. The average horizontal kinetic energy produces an average velocity of 2.2 m/s (5 mph) or about 3 PW over the planet[8]. To be useable the actual wind velocity must be in the technically useful range over an inhabited land area during a period that matches a demand for power.

Table 9.4 shows the relation between turbine blade length, average wind power, and electric power output. The wind power increases as the cube of the linear speed. The advantage of scale increases by approximately the blade length to the 2.4 power. Wind speeds that range over a factor-of-two therefore produce power that ranges over a factor-of-eight. This limits the useful range of most wind turbines to about a factor-of-two range of wind velocity.

Table 9.4 Wind power vs turbine rating

Rated power, kW ---	1250	625	313	156	78
Diameter, m ---------	100	50	25	13	6
m/s mph W/m2 --------- total wind kW ----------					
5 11 73	572	143	36	9	2
6 14 126	988	247	62	15	4
7 16 200	1569	392	98	25	6
8 18 298	2342	585	146	37	9
9 20 425	3334	834	208	52	13

Wind turbine technology reached a high state of the art using fluid dynamics developed to design aircraft[9]. Modern wind turbines are three blade propellers on a horizontal hub at a height roughly equal to the blade diameter. The blade angle determines the fraction of the wind-to-rotational energy efficiency. The optimum energy transfer depends on the speed of the blade, the angle between the blade and the wind, and the speed of the wind. The blade twist matches the angle to the speed at different diameters.

Wind-to-electric power conversion is a maximum at a blade/ wind speed ratio of about 10 at the blade tip. Adjustable speed propellers rotate to optimize the angle at different wind speeds. A well designed propeller recovers about 50% of the wind energy passing through it.

The hub contains the generator, power conditioning, pitch and yaw adjustment mechanisms, and mechanisms to detect and prevent a variety of problems associated with high, gusty winds. These include aerodynamic stalls that occur from sudden changes in wind speed and/or direction as well as mechanical failures that intensify as towers become higher and turbines become heavier. All of these mechanisms must function automatically.

The size of individual units is mostly a compromise between the economies of scale, instability, and cost effectiveness of extremely large structures. The most economical units for many wind regions are up to 1000 kW.

In areas that are suitable for commercial scale wind power a *wind farm* of many modules is necessary to develop full potential of the resource. Both the number of modules in the operating unit and the characteristic of the wind at the particular site affect the cost effective module size. The cubic dependence of wind power on velocity limits the acceptable wind velocity to a narrow range of about 7-10 m/sec or 15-25 mph.

The wind history of many regions is now mapped continuously[10]. Sites suitable for exploitation have average wind velocities of 7-10 m/s at least 60% of the time or 5000 hrs per year. The northern coasts of Europe and the central plains of the U.S. are particularly favorable.

The most reliable sites have a long stretch of ocean or flat terrain facing the prevailing westerly winds generated by the general flow from the tropics to the poles. On the seacoast there is an added effect from the daily differential land-ocean heating and cooling. Further inland, sites along ridges that face the prevailing wind produce winds that are not affected by ground turbulence.

Table 9.5 shows that five nations accounted for most of the wind power installations in 2000[11]. This was a breakout period for wind power expansion so these numbers are temporary.

Table 9.5 Wind power of major users

	wind MW	total GW	percent wind
Germany	6107	65	9.5
United States	2610	457	0.6
Spain	2836	25	11.3
Denmark	2341	4	56.7
India	1220	62	2.0
Netherlands	473	10	4.6
Italy	425	31	1.4
United Kingdom	424	43	1.0
China	352	155	0.2
Sweden	265	17	1.6
World	18449	1752	1.1

Table 9.6 shows the allowable capital cost for two assumptions about the average price the power can command.

Table 9.6 Allowable capital cost for wind power

Nameplate power	1 MW
Price	$0.06 /kWh
Utilization`	0.60
Revenue	$315 /y thousand
Maintenance	$20 /y thousand
Payment on capital	$295 thousand
Amortization	25 y
Return on investment	0.08
Allowable capital cost	$3,687 thousand

Wind power is a new technology that has broken through the development cost barrier and has begun to acquire operating experience to be competitive on a broader scale. It requires unobstructed terrain, but is compatible with dual use of land. It has potential to supply a major fraction of the power to restricted locations, but not the reliability for the power grid. It could be a good source of power for electrolytic hydrogen for transportation.

Biomass Power

Biomass is the solar power stored by plant life plus the combustible residue of organic matter produced by contemporary photosynthesis. Three centuries ago, when the world population was a hundred-fold smaller, forests were plentiful and wood was the primary energy source. Most of the energy was used for cooking and heating. At least half the world population still relies on biomass for cooking and heating. Meanwhile the developed world looks at biomass as an alternative energy source for transportation. There are problems with the arithmetic of this view.

Table 9.7 summarizes the plant growth distribution of carbon deduced from satellite photography which divides the Earth into a grid at 0.5° intervals of latitude and longitude. Each element of the grid is assigned to one of 32 different classes of vegetation[13]. Most of the Earth's area is oceans, polar ice, and deserts which support little or no vegetation. Arable land is mostly short lived vegetation and crops that are planted annually. The carbon in forest lands is stored as trees. Trees contain more carbon than food crops but their growth typically covers lifetimes of 20 to 80 years. The productivity of a growing timber tree increases for at least the first twenty years.

Table 9.7 Inventory of the Earth's plant carbon

	% of area	% of carbon
Unusable area	82.6	3.4
Arable land	10.9	26.7
Forest land	6.5	69.9
Total plant carbon	606 Gt	
Carbon heat of combustion	9106 kWh/t	
Energy in live plants	5518 PWh	
Storage rate	0.186 PW	

The effective rate at which photosynthesis currently stores solar energy is estimated by adding the arable land growth to the average forest growth divided by a 25 year growth period. The resulting 0.186 PW that was shown in Table 9.1 is about 10 times the present energy consumption by humans.

169

The total biomass is the 606 Gt of carbon in living plants plus the accumulated mass of vegetation that is dead and rotting. The dead and decaying matter obviously varies widely and is not easy to determine worldwide. The total that was used in Fig. 6.8 is 3.5 times the live plant biomass.

Table 9.8 is a calculation to place these data in perspective. It is an estimate of the efficiency of a rainforest of the Pacific Northwest in storing solar energy as a Douglas fir log. According to these values about 0.1% of the solar energy reaching a forest which is mostly obscured by clouds is converted by photosynthesis to the main logs of the trees. The 11.7 km/m^2 carbon content calculated for this log is consistent with the maximum carbon content of conifers of the Oak Ridge model if 60% of the carbon content is in the main log.

Table 9.8 Estimated rainforest carbon production

Tree area	100 m2/tree
Solar exposure	4 h/day
Solar energy per year	1460 kWh/m2
Average log diameter	0.50 m
Average log length	30 m
Average log density	500 kg/m3
Total log mass	2945 kg
Fraction carbon	0.40
Carbon yield	11.7 kg/m2 (in log)
Heat of combustion	9.1 kWh/kg
Average growth period	25 y
Photosynthesis efficiency	0.0029

Forest biomass is now harvested for paper, lumber and related forest products. A portion of this land is in tree farms which reproduce forests on a continuing cycle. Historically forests have been cleared to produce even more valuable agricultural crops on the underlying arable land. The land that is now being farmed is the land most suited to agriculture. It is probably the only land that ought to be farmed[14]. Further deforestation has diminishing return for agriculture and reduces the diversity of essential plant, animal, insect, and microbial life with consequences that are unforeseeable.

The literature of biomass technology covers most of plant biology, agriculture, and forestry[15]. The emphasis of the associated chemical technology is on sustainable alternatives to petroleum as portable liquid fuel for transportation. Methanol and ethanol are the liquids that are easiest to produce. They have about half the energy hydrocarbons. To be used as a stand alone fuel they would require new engine technology.

There is a major problem in thinking of biomass as the alternative to the massive present demand for petroleum based energy. Much of the arable land suitable for producing biomass fuels already has a high priority use, *food*. Many parts of the world must already grow several crops a year to feed their population. The bio-diesel, methanol, and hydrogen alternatives all require additional energy cost to produce. Economic competition for biomass energy for transportation and biomass food production must inevitably exacerbate the existing conflict in agricultural social policy.

The 18 GW of electric power from biomass is now produced mostly as co-generation from municipal and industrial waste disposal. This use should continue, but it cannot grow to more than a minor fraction of the total demand.

Geothermal Power

The average heat loss through the Earth's crust is about 0.1 PW after cooling for 4.5 billion years[16]. The surface temperature balances 121 PW of solar power input, 0.1 PW of heat from the core, minus thermal radiation. At the surface the temperature changes in daily and seasonal cycles. At a depth of 5-10 m the temperature is constant at close to the annual average maximum and minimum surface temperature. Below this the temperature increases with depth until it reaches a practical limit for drilling exploration at about 3 km. Geothermal power applications are designed to take advantage of the constant temperature at shallow depths, fortuitous hot spots at intermediate depth, or deep sites.

Heat pump geothermal technology uses a heat engine operating in reverse to take advantage of the constant temperature at shallow depths. Instead of expanding gas to produce power a heat pump uses power to heat a gas by compressing it. Depending on whether the temperature requires heat or cooling the heat pump can be connected to transfer heat either to or from the reservoir[17].

An underground site is a suitable heat reservoir if it has a constant temperature between the extreme annual surface temperatures with enough ground water flow to redistribute the heat. T-S diagram in Fig. 7.2 showed that if the temperature difference is small a large quantity of heat can produce only a small quantity of work. If the arrows are reversed the heat engine becomes a heat pump. Performing work on the gas now transfers a large quantity of heat. The heat pump becomes more efficient under the conditions that make an engine less efficient.

Wet steam technology recovers heat and power at sites where steam percolates upward from an aquifer that has contacted a thermal source. Hot spots in the lithosphere have two main sources. Some sites originate with a pool of hot magma that is created at shallower depths by the energy of plate tectonic collisions. Others are associated with plumes of magma coming more directly from the Earth's core. These sources are not easily distinguished. They are a part of the circulation of the magma[18].

Most of the nearly 60 GW of wet steam geothermal power is consumed for residential heating and industrial process heat. Hot water at near the normal boiling point would have to use binary Rankine engine technology to produce electric power.

Wet steam geothermal sites are developed further for commercial electric power by drilling to depths that have pressures up to 500 psi (34 bar). This is well below the pressure used for high performance steam turbines, but much more efficient than hot water. About 4.8 GW of electric power is being generated worldwide, particularly in Iceland and New Zealand. A site in Northern California known as *The Geysers* is a cautionary note about overestimating the potential of a site[19]. At least 14 different generating units provide a total of 1 to 2 GW. Potential wet steam geothermal sites with up to perhaps 15 GW have been identified.

Hot dry rock geothermal power has potential wherever the Earth's crust is thin enough to reach a high temperature within about 3 km of the surface. Heat is extracted by injecting water into the site to produce high pressure steam. Although it has wide geographic application, it is not unlimited[20]. The rock must be porous and/ or easily fractured by explosives. The total heat is limited by the thermal conductivity of the rock that separates it from the core. The sensible heat due to the heat capacity of the rock at the site provides an initial burst of energy. Once that heat is used, replenishing it by thermal conductivity from the core is slow. The recoverable heat may be limited to the sensible heat of the rock that can be exposed. Like a dry oil well, the rate of heat flow is eventually too slow to support the plant.

Geothermal power is not a renewable supply of free energy in the same sense as ocean, solar, or wind power. It is a resource to be discovered, developed, exploited, and depleted like a mineral deposit. The thermal power extracted at a particular site is limited by heat flow to the well head, like the flow of petroleum in an oil well. The site must be evaluated and developed as a resource that will diminish with use.

The price of hot dry rock power includes the cumulative cost of exploration, development of access to hot rock, and production of power. The development includes fracturing the rock of the site to give access to a large volume that can be filled with high temperature, high pressure steam. Each site is a unique development and production problem. Heat flow to the well head through rock with low thermal conductivity is the problem to be solved.

Geothermal sites may or may not be environmentally clean, depending on the gases and brine associated with the source. Wet steam does not greatly change a hazard that already exists.

Ocean Tidal Power

The origin of ocean tidal power is gravitational attraction by the Moon with a smaller contribution by the Sun. The tidal period coincides with the twice daily passage of the Moon due to the rotation of the Earth with a phase lag. The amplitude of the daily tide with the rotation of the Earth is modulated by the rotation of the Moon about the earth with a period equal to the lunar month. This is further modulated by the twice yearly alignment with the gravitational force of the Sun which has a magnitude at the Earth about half that of the Moon[21].

The total magnitude of tidal power is roughly 1 PW. This is based on the 2 msec/century increase in the length of the day due to tidal friction. This information is from the observed growth rate of cockle shells[22].

The magnitude is best explained by a model that treats each ocean as an oscillator with water sloshing back and forth at a resonant frequency with harmonics related to the shoreline terrain. It dissipates energy by friction at the shorelines and bottom, but otherwise conserves energy. The power is derived mostly from tidal harmonics with periodic additions by the gravitational forces plus smaller increments from the prevailing winds.

The cost of tidal power and the method used to extract it depend on the characteristics of the site. The differences are mainly in the structures needed to concentrate the flow at the turbines. Well designed turbines extract 85% of the energy. The energy is highly predictable but varies widely, is available only 30 to 50% of the time, and has no relation to the timing of the demand.

Dam and weir technology is applicable at sites with abnormally large tides flowing through a restricted channel to a large tidal basin. The dam increases the linear velocity by limiting the flow cross section. Additional dams may be required to block other entries to the basin. A reversible reaction turbine takes energy from both the incoming and outgoing tides. The power is proportional to the head.

An example is a 250 MW installation on the Rance Estuary near Mont Saint-Michel, France. Most other actual installations are in the 10-50 MW range. The extraordinary tides in the Bay of Fundy in northeast Canada have made it the object of many studies of the potential of tidal hydroelectric power.

Dam and weir technology requires extremely large water flows. The applications are obviously limited to only a few sites in specific locations.

Open ocean technology uses a farm of free standing turbines anchored to the ocean bottom analogous to wind turbines. This is applicable to a much greater number of sites with individual units of more modest size. They are more amenable to a distributed source philosophy of grid organization. The energy density of water flowing at 3 miles per hour is 20 times smaller than that of a 2 meter head behind a conventional hydroelectric source.

The random timing of tidal flow with respect to the demand for electric power is a major drawback to tidal power. The magnitude and timing of the power are predictable with precision and reliability. However the timing of the peak tidal power continuously cycles in and out of phase with the peak demand.

Small scale units providing enough power to illuminate a navigation buoy have been constructed using the vertical oscillation of the buoy with respect to its mooring post. The head provided by waves is even smaller than that by tidal flow.

Ocean Thermal Power

The top 100 meters of warm tropical oceans is one of the Earth's major heat sinks. It holds about 3500 PWh of heat at a temperature 20-26 K higher than the deep ocean temperature at an energy density of 2.5 kWh/t. This is equivalent to the energy in a 1-meter head of water in a hydroelectric or tidal system. The temperature difference can be used to drive a heat engine.

Practical public power is restricted to tropical coastline locations. The 100 MW unit used for illustration is probably necessary for economy of scale but may be a practical limit for the scale of Rankine cycle technology based on refrigerant gas. Since the seasonal cycle of tropical oceans is small the demand and price would be suitable for base power.

The cold water is extracted below the boundary region between the *surface ocean* and *deep ocean*. The cold water duct must reach a depth of about 1000 m. It could be a concrete structure supported by the slope of a moderately steep coastline. The geology of the coastline is a significant factor in choosing the site.

The quantity of water is large enough to have a significant effect on local marine ecology. The cold water discharge would be beneficial to marine organisms. The normal mixing rate of the Deep Ocean is measured in centuries, if not millennia. It depends on specific cold water streams in ways that are only partly understood. The environmental effect of large scale exchanges of warm and cold water must be considered carefully.

A United Nations commission identified 80 nations having coastlines with qualifications for ocean thermal power[23]. Typical candidates might be nations like the Federated States of Micronesia with tropical island coastline totaling 6100 km, almost as much as India. The individual islands are not large enough to support major power plants. An ocean thermal power plant can be designed to suit the capabilities of the site and demand at a particular location.

Ocean thermal power is not seriously influenced by seasons, time of day, or weather. It could serve as back-up power for less reliable distributed sources or as an all purpose power source for an isolated area with use equivalent to a population of 100,000.

Fig. 9.2 is a schematic diagram of a system using the surface and deep ocean waters as hot and cold reservoirs of a Rankine cycle heat engine. It uses ammonia as an operating fluid that vaporizes at the temperature of the warm surface ocean and condenses at the cold temperature[24]. One pump circulates warm surface water through a heat exchanger that vaporizes the liquid ammonia and heats the gas to a temperature and pressure that can drive a turbine to generate electric power. A second pump circulates cold deep ocean water through a second heat exchanger which cools the ammonia, condensing it to liquid. A third pump injects the liquid ammonia into the warm water heat exchanger, where it forms ammonia vapor at a high pressure.

The temperature-entropy diagram shows the low efficiency, about 5%, that follows from the small temperature difference. This is 1/6 that of a conventional thermal power plant and nearly 1/20 that of the highly efficient engine cycle of a hydroelectric plant. The temperature difference that transfers the heat is divided between the ammonia and water. This fraction of the available heat extracted from the water gives an overall efficiency of less than 1%.

The 30 meter diameter of the sea water ducts illustrates the scale. The trade-off between the cost of pumping energy and the size of the ducts has a narrow range. The flow kinetic energy rises sharply when it becomes significant since it is proportional to the square of the linear flow rate. This occurs at about ¼ meter per second.

The water pumps need only supply flow kinetic energy to the balanced hydrostatic pressure of the intake and outflow. The flow rate for the liquid ammonia compressor and gas turbine is about 15 tons per second. Ocean thermal power is not mature technology and would require development for both machines and materials that would justify the long amortization period.

There is nominally no environmental emission. The turbine engine uses potentially hazardous ammonia gas at high pressure in a sealed system. The quantity of ammonia is large enough to conflict with the competitive demand for use as fertilizer. The cold water duct requires deep sea construction with disruption of the path to the intake. The length of the path depends on the steepness of the slope to a depth of about 1 km. Protection from tropical storm damage may be an issue.

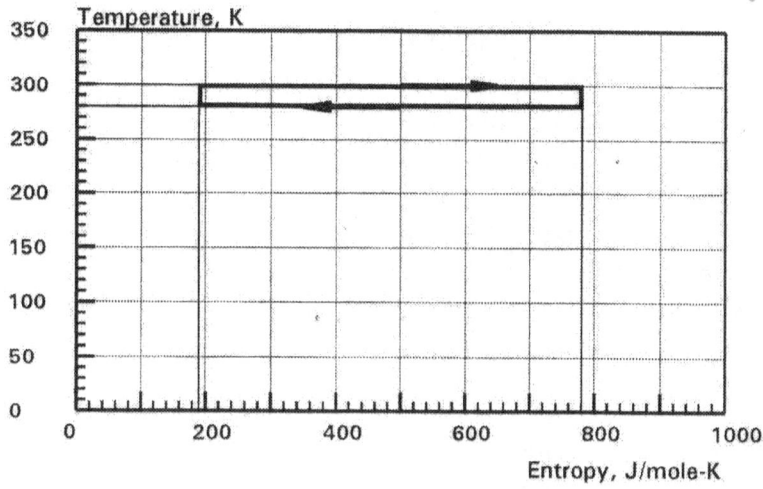

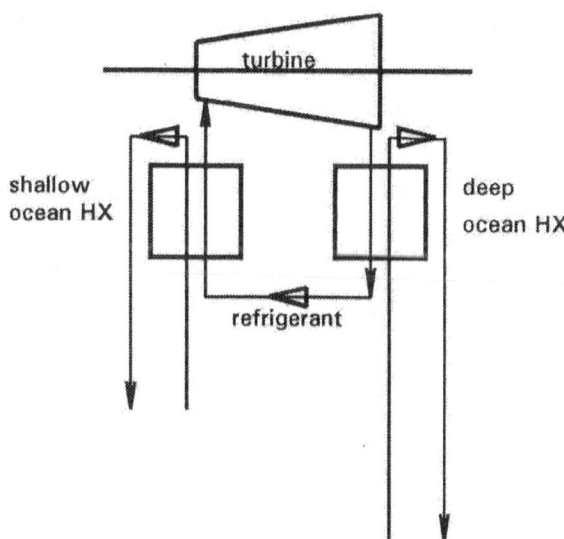

Fig. 9.2 Schematic ocean thermal power system

179

Table 9.9 correlates the revenue with the allowable cost of a plant to deliver 100 MW of useful electric power. The flow rates are for a temperature drop of 20 K across the generator turbine and a 2 K drop in each of the sea water-ammonia heat exchangers.

The absence of fuel cost allows low operating cost and high capital cost. At $4100/kW the allowable capital cost would be over two and a half times the cost of a coal fired plant. The distribution of capital costs that is shown is obviously dependent on the site and trade-offs that are yet to be determined by the development. The high utilization assumes that commercial applications would take advantage of time-of-use pricing.

Table 9.9 Economics of a 100 MW Ocean Thermal Power Plant

Flow specifications

Nameplate power	100 MW
Heat engine efficiency	.07 W_e / W_h
Maximum heat input	1483 MW
Water volume flow rate	182 m3/s (2 channel)
Water duct diameter	30 m
Water pump power	12.1 MW (2 pumps)
Liquid ammonia volume flow rate	15 m3/s
Ammonia pump power	19.7 MW

Revenues

Plant utilization	.90
Annual energy production	788 GWh
Price	$.06
Annual revenue	$47.3 million

Cost allocation

Annual operating cost	$6 million
Annual payment on capital	$41.3 million
Amortization period	25 y
Return on investment	.12
Allowable plant cost	$413 million

Fuel Cell Energy Storage

The rising price of gasoline is an incentive to look for alternative ways to store energy in portable form. The advanced state of the technology and a prospect that hydrogen might replace fossil fuel for transportation motivates interest in hydrogen fuel cells. A massive new industry would produce electrolytic hydrogen to store off-peak power for transportation and for use during peak demand,

The manned space program developed a practical alkaline hydrogen fuel cell for electrical power. Commercial units are now marketed as a 100-500 kW substitute for diesel-electric back-up power. Gasoline at $2.35/gallon produces the same power as a fuel cell operating on hydrogen produced by electrolysis at $0.12 /kWh[25].

The fuel cell principle is similar to a rechargeable battery[26]. *In the galvanic or battery mode* hydrogen gas reacts to produce electrons at the positive anode half cell. Oxygen gas reacts to remove electrons from the negative cathode half cell.

$$H_2 + 2\ OH^- \rightarrow 2\ H_2O + 2\ e^- \text{ (anode)}$$

$$\tfrac{1}{2}\ O_2 + H_2O + 2\ e^- \rightarrow 2\ OH^- \text{(cathode)}$$

Electrons flow from anode to cathode in the external circuit. The H^+ ions flow from the anode to the cathode through a membrane that separates the gases.

In the electrolysis mode an external voltage is applied to reverse the polarity of the cell. This reverses the current flows, the anode-cathode designation, and the cell reactions. It stores energy in the form of hydrogen and oxygen gases.

The proton exchange membrane that separates the reagent gases is a polyelectrolyte polymer with multiple hydrogen bonds that produce the high ion conductance by a cooperative shift mechanism[27]. When a proton is absorbed at the negative side of the membrane a concerted shift in hydrogen bonds of the membrane immediately releases a proton at the other side of the membrane. A similar mechanism gives protons exceptionally high equivalent conductivity water solutions.

Fig. 9.3 is a magnified cross section of a fuel cell stack. Porous electrodes sandwich the proton exchange membrane in the center of each cell. These are in turn sandwiched between metal terminals that support the electrode-membrane structure mechanically and connect electrically to the porous electrodes. Both the electrolyte and appropriate gas reagent circulate in channels which provide maximum contact with the porous electrode.

The metal spacers form a bipolar electrical connection between adjacent cells. The anode of one cell is on one side. The cathode of the adjacent cell is on the other side. This connects a stack of n cells in series with a common current and a total voltage n times the single cell voltage.

The fuel cell efficiency is limited by *over-voltages,* internal resistances that reduce the external cell voltage and dissipate the energy as heat. Electrochemical kinetic properties of each boundary and each medium make contributions that increase with increasing current. The electrodes are typically carbon fiber felt which permit both free gas flow and high current density. They give both the reagent gases and ions in the electrolyte solution an electrical contact with the membrane. The alkaline hydrogen cell produces reasonably small over-voltage without precious metal catalysts. A balance between the porosity and surface tension of the electrode accommodates both gas and the electrolyte.

Fuel cells are not yet a robust technology with proven long term reliability. The assembly of components is expensive and complex, but straightforward. Once assembled, it may not be easy to maintain and replace. It remains to be determined whether it can be made robust enough to warrant an amortization period long enough to finance it. Apart from the potential explosion hazard and large gas storage requirement the fuel cell storage attributes are environmentally favorable.

In principle, a fuel cell can convert the energy of any spontaneous reaction directly to electric power by electrolysis[28]. By reversing the cell current it can produce the high energy reactants. In practice, economical large scale power requires a cell designed for each specific reaction. This discussion focuses on energy storage by alkaline-hydrogen cells that use proton exchange membranes.

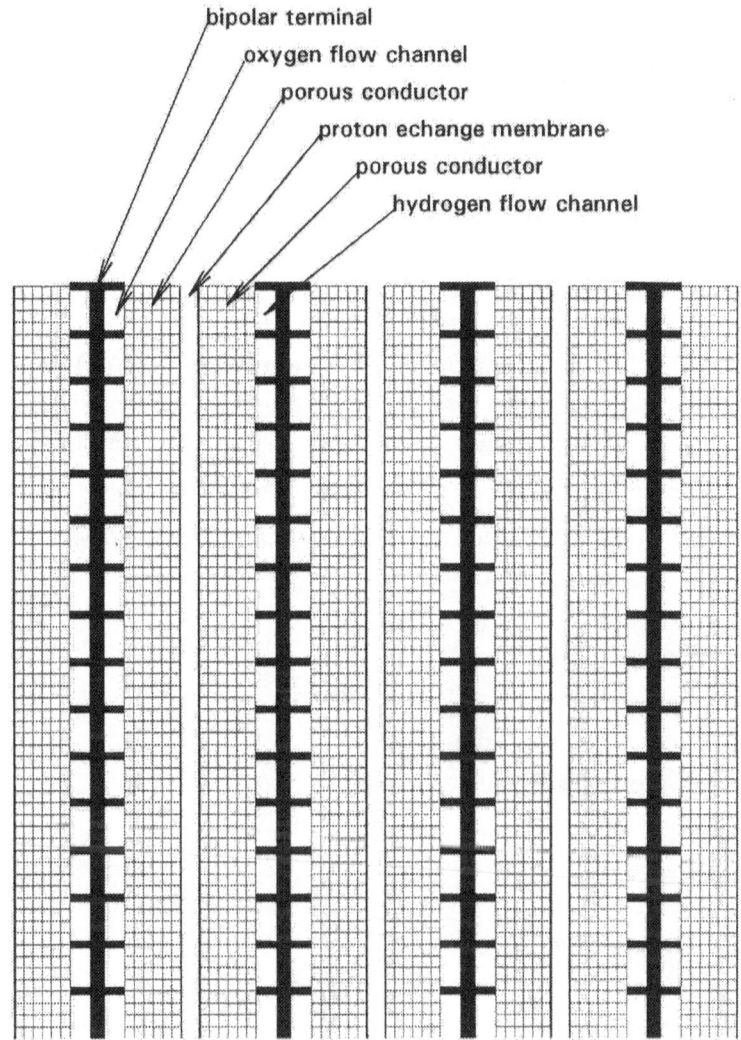

bipolar terminal
oxygen flow channel
porous conductor
proton echange membrane
porous conductor
hydrogen flow channel

Fig. 9.3 Schematic elements of a hydrogen fuel cell
Cross section of parallel cell assembly

The chemical-to-electrical efficiency approaches one as the current approaches zero. The voltage approaches the free energy of reaction in electron-volts. A practical cell must operate at high current density with low voltage loss. The cell stores low value off-peak electrical power as hydrogen and oxygen gases. The electrical power is recovered at higher value by the reverse reaction forming water.

A 200 MW unit could store the 1200 MWh or so of energy required for daily peaking power of a 1 GW plant. The electrolysis system requires at least 25,000 m² of membrane surface in stacks of 100 or more individual cells.

Table 9.10 summarizes the economics of a fuel cell storage facility that stores low value power from a base power plant as hydrogen and oxygen gases as fuel for diverse applications that include transportation as well as peak demand power.

Table 9.10 Hydrogen fuel cell storage economics, 200 MW

Cell characteristics

cell voltage	0.8 volts
current density	2000 amps/m²
power density	1.60 kW/m²
total membrane area	125,000 m²
hydrogen gas, 12 hr @100 bar	4,000 m³

Revenues

nameplate power	200 MW
utilization	0.50
storage cost	0.06 $/kWh
annual electric energy	876 GWh/yr
annual revenue	$52.6 millions/yr

Operating cost $ 17.5 millions/yr

Capital cost

annual payment on capital	$ 35.0 millions/yr
rate of return on investment	0.08
amortization period	25 y
allowable capital cost	$437 million

Light weight cylinders wound from carbon filament fibers can store hydrogen at 5000 psi. It costs 10-15% of the energy to compress the gas. The outlet pressure reduction valve would be inside the tank. A cylinder strong enough to hold hydrogen could withstand the impact of accidents that are reasonable to expect. Refueling could use exchangeable tanks.

Interregional Power Transmission

A major regional grid has numerous connections with adjacent regions[29]. However the perception that long distance transmission lines can solve reliability problems by connecting a greater diversity of power sources has several dangers. The grid is a network of local and regional distribution centers. Most of the power connects with plants within a radius of 100-200 miles. Exceptions such as the Hoover Dam transmission lines transport power from a particular source to specific distribution points. Each distribution center sets the voltage at which it accepts power. The power between regions flows in whatever direction local voltages dictate. The transmission line operator learns the identity of the buyer and seller only after the fact.

Power failures propagate downward through the grid hierarchy with much more certainty than upward. Transmission lines increase the vulnerability to system wide failure. They are least reliable when the demand is highest. The power must be predictable and well within the capacity of the lines. It is to be viewed as base power to be scheduled well in advance, not as peaking power or back-up for emergencies.

A power failure or sudden change in load anywhere in the system causes a transient oscillation which may or may not be damped. This occurs much too fast for analysis and response by a human operator[30]. A human response can either isolate or intensify it. The transient may trigger an automatic protective response by circuit breakers. This can cause a failure to propagate to adjacent systems.

Notes and references for Chapter 9

1. The values in this table are similar to those in Richard C. Dorf, The Energy Factbook, McGraw-Hill, New York, 1981. They are consistent with other data, such as the satellite data cited in Jae Edmonds, and John M. Reilly, Global Energy, *Assessing the Future*, Oxford University Press, 1985.

2. The radiation temperature of the solar input is 364 K over an area πr^2. This would be balanced by emission from a black body at 278 K over an area $4\pi r^2$.

3. The hydrostatic pressure of water at a generator is ρhg Pascal where, ρ is the density in kg/m^3, h is the head in m, and **g** is the gravitational acceleration, 9.8 m/s^2. The gravitational potential energy produces maximum flow kinetic energy, $\rho hg\ A_f\ v_x$ watts, at flow cross section, A_f, and linear flow rate, v_x m/s.

4. The additions are compiled from anecdotal summaries contained in the Annual Energy Review 2000 of the Energy Information Agency of the U.S. Department of Energy.

5. Thomas G. Alexander, *The Grand Coulee Dam, the Columbia River, and the Generation of Modern Washington*, a chapter in Politics in the Postwar American West, Richard Lowitt, Ed. University of Oklahoma Press, 1995

6. An account of the people, history, and economics of the area affected by the Three Gorges Dam project, Deirdre Chetham, *Before the Deluge*, Palgrave MacMillan, New York, 2002

7. The report of a 1968 commission of the United Nations lists 20 existing and 52 projected pure pumped storage facilities in the U.S. and Europe. About half have capacities greater than 0.3 GW. About half operate on a daily cycle in addition to the seasonal cycle. *The Future Role of Pumped Storage Schemes for Peak Load Hydroelectric Supply*, Department of International and Social Affairs, United Nations, 1968

8. A friction cross section, A_f = 5.1 x 10^{14} m^2, the area of the earth, an average wind speed, v_x = 2.2 m/s (5 mph), and air density ρ = 1.16 kg/m^3 give total power P = ½ $\rho\ v_x^3\ A_f$ = 3.0 PW.

9. E.A. DeMeo, and P. Steitz, *Wind Power*, in Advances in Solar Energy, **6**, 1990

10. David Milborrow, Andrew Garrad, and Birger Madsen, *Wind Energy, the Facts*, European Wind Energy Association, 26 Spring Street, London, U.K. 1999

11. Edward S. Cassedy, *Prospects for Sustainable Energy*, Cambridge University Press, Cambridge, U.K., 2000

12. J. Beurskins, and P.H. Jensen, *Wind Energy*, Chapter 6 in *The Future for Renewable Energy*, EUREC Agency, James & James, London, 2002

13. J.S. Olson, et.al., *Major World Ecosystem Complexes Ranked by Carbon in Live Vegetation: A Database*, Oak Ridge National Laboratory, Carbon Dioxide Information Analysis Center, 2001

14. Dennis Avery, Hudson Institute Report, 1998

15. The proceedings of the 1998 international conference on *Biomass for Energy and Industry* sponsored by the Commission of European Communities at

Wurzburg, Germany is a comprehensive biomass technology reference containing more than 600 peer reviewed technical papers each containing further references. CARMEN, Rimpar, Germany, 1998

16. The present rate of heat loss from the Earth's core by thermal conductivity is 0.1 PW based on a thermal conductivity coefficient 1 W/mK, earth area 5.1 x 10^8 km^2, mean crust thickness 3.8 km, and magma temperature 1100 K.

17. W.A., Duffield, J.H. Sass, and M.L Sorey, *Tapping the Earth's Natural Heat*, U.S. Geological Survey Circular 1125, 1994

18. G.R. Foulger, and J.H. Natland, *Is Hotspot Vulcanism a Consequence of Plate Tectonics?*, Science, 300, 921, 2003

19. 2 GW of electric power capacity costing $3.5 billion was installed at sites in an extinct volcanic region known as *The Geysers*. They were expected to produce 6% of the power requirement of California indefinitely. After 6 years productivity was at 1.5 GW and falling. Richard A. Kerr, *Geothermal Tragedy of the Commons*, Science, **253**, 134, 1991

20. J.W.Tester, and H.J. Herzog, *Economic Predictions for Heat Mining: A Review and Analysis of Hot Dry Rock Geothermal Energy Technology*, Massachusetts Institute of Technology, 1990

21. Gravitational acceleration is proportional to mass and inversely proportional to distance from the center of mass cubed. The values are; Earth, 9.8 m/s^2, Moon,.00047 mm/s^2, 20 million times smaller, and Sun, .00021 mm/s^2.

22. Cross sections of cockleshells that were periodically submerged and exposed by tides have a, daily, monthly, and the annual modulation. A comparison of cockleshells from different geological ages indicates that the Earth day increases by 0.002 sec/century. This corresponds to a frictional energy of 1000 TW. Note that friction this large cannot possibly have been constant. The increased friction may be due to the proliferation of oceans and shorelines by plate tectonics. T.Ohno, in *Tidal Friction and the Earth's Rotation,* P. Brosche, and J.Sundermann, eds., Springer-Verlag, Berlin 1982

23. *A Guide to Ocean Thermal Energy Conversion for Developing Countries*, Department of International and Social Affairs, United Nations 1984

24. Refrigerants are gases at ambient temperature and pressure that liquefy at high pressure. This qualifies them for use in a Rankine engine cycle at temperatures in the ambient temperature range. Their principle application is in cooling by a reverse Rankine cycle.

25. Assuming 0.80 efficiency for each electrolysis step and 0.33 efficiency for the gasoline engine

26. The early patents on gas phase electrochemical cells were issued to William R. Grove in 1839.

27. Proton exchange membranes are ionomer polymers which conduct a current by rapid exchange of protons. Dupont nafion is a teflon base with sulfonic acid groups. The polymer is a strongly hydrophilic proton donor.

28. Gregor Hoogers, Ed., *Fuel Cell Technology Handbook*, CRC Press LLC, 2003

29. Connections of the New England Power Pool to lower cost Canadian power involve more than 488 generators, 816 load centers, and 2,11/7 power buses. Robert T. Eynon, Thomas J. Leckey and Douglas R. Hale, *The Electric Transmission Network: A Multiregional Analysis*, Energy Information Agency Report, U.S. Department of Energy

30. The grid's normal reaction to a sudden change in demand is a damped oscillation about the ultimate steady state. Glitches that decay in less than about 300 milliseconds (18 cycles) make a transition to a new voltage. Glitches lasting longer may require a higher-level response. IEEE Power and Energy, **1**, 36, 2003

Chapter 10
Solar Power Principles

The maximum solar radiant power at the Earth's surface is typically 1 kW/m² over a broad area. A black body absorber that absorbs it directly can reach a maximum temperature of 364 K or 91 C. Optical concentration can increase the temperature up to a theoretical maximum equal to the surface temperature of the Sun, about 6000 K. This distinguishes the two classes of solar power technology.

Photovoltaic power systems rely on a version of the *photoelectric effect*. When a substrate absorbs a photon with energy greater than the critical threshold energy for that substance, the substrate ejects an electron. In a vacuum photocell the substrate is the cathode that emits a flow of photoelectrons to the anode of a circuit. Photovoltaic systems are non-concentrating semiconductor panels that produce the same effect over a large area. The voltage is the threshold energy in electron volts. The current is the flow of electron charge per second.

Solar thermal systems concentrate solar radiant power with optical mirrors and convert the energy to heat in a heat engine working fluid. The essential characteristic of the collector optics is their *concentration ratio*, a decrease in radiant power cross section that increases the effective temperature the radiation can produce.

Solar power that is reliable is limited to arid desert locations within transmission line distance of population centers. The clear sky solar power can be predicted at any specific latitude, day of the year, and time of day. Weather makes the actual radiant power unpredictable at most locations. United States, Mexico, Brazil, Argentina, China, India, Egypt, Pakistan, Turkey, and nations of Northern Africa and the Middle East have significant useful resources. This is comparable to the quantity of useable hydroelectric power.

Solar Radiation Cycles

Fig. 10.1 shows the tilt of rotational axis of the Earth in relation to the position of the Earth in its orbit around the sun. The solar orbit is a nearly circular ellipse. The Earth's polar axis tilts at an angle of 23.45° with respect to the plane of the solar orbit. Conservation of angular momentum fixes a constant tilt direction in space. Vectors at various days of the year show the direction and magnitude of the radial and transverse components of the tilt with respect to the direction of the Sun. The radial component is responsible for the seasons. In the northern hemisphere the tilt away from the Sun is a maximum at the *winter solstice,* December 22. The maximum tilt toward the Sun is at the *summer solstice*, June 22. At the *spring equinox*, March 22, and *fall equinox*, September 23, there is no radial component.

Fig. 10.2 shows the Earth in relation to the Sun. Planes perpendicular to the rotational axis show the equator and a site at 35° north latitude. A plane at the angle of the tilt component divides the 35° north latitude plane into day and night. The figure shows the 23.45° tilt angle at the winter solstice. The heavy section of the 35° north latitude plane is the daylight path of a site. This path is shorter than a semicircle in winter and longer than a semicircle in summer. The sites progress from right to left, dawn to noon. Each solar ray intercepts a local vertical through the center of the Earth at the site in question. The path length through the atmosphere is inversely proportional to the cosine of the angle between the Sun and the local vertical. This compound angle changes with daily rotation as well as the seasons. At an equinox the day is 12 hours long from 6 a.m. to 6 p.m. solar time regardless of latitude.

At 35° north latitude the day shortens to 9.6 hours minimum at the winter solstice and lengthens to 14.4 hours maximum at the summer solstice. At latitudes within the tilt angle from either pole a fraction of the year is in darkness with no sunrise. In summer the same fraction of the year is in daylight with no sunset.

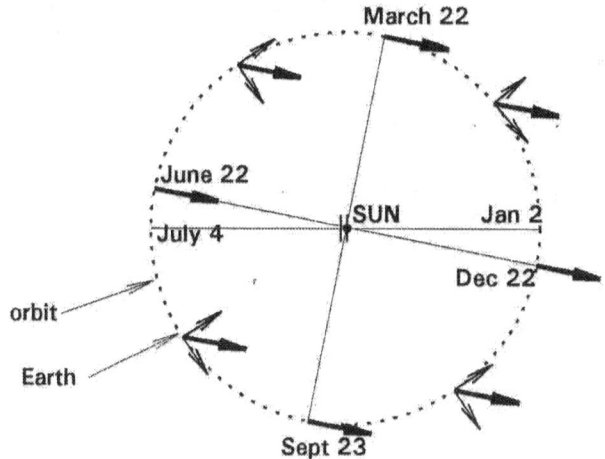

Fig. 10.1 Earth's orbit showing components of the tilt of the north pole toward and away from the sun

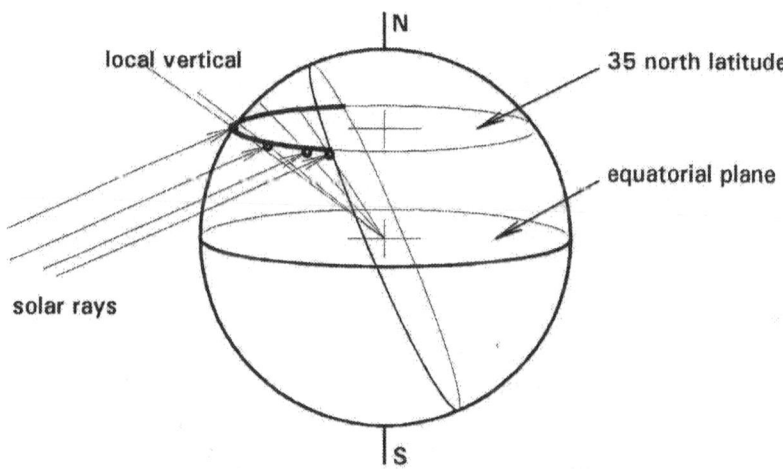

Fig. 10.2. Local coordinates at 35 degrees north latitude in relation to illumination at the winter solstice

Fig. 10.3 shows the daily solar radiant power through a clear sky atmosphere at 35° north latitude. The power has components that depend on the effect of the changing angle of the sun on the absorption path length through the sky. The following steps yield the total radiant power as a function of latitude, day of the year, and time of day.

Effective spectral absorption coefficients for 40.6° north latitude follow from the ratio of the two spectra that were shown in Fig. 6.1. These data are for noon at an equinox at the National Renewable Energy Laboratory at Boulder, Colorado. Reference spectral absorption coefficients at each wavelength for the equator at noon of an equinox are calculated by correcting the path length for the difference in latitude.

The path length for latitudes other than the equator require adjustment for the radial component of the tilt at the particular day of the year, and the east-west angle of the sun from the local vertical at each time of day to obtain spectral absorption coefficients at the particular latitude, time of day, and time of year. [1]. Integrating the spectral radiant power over wavelength gives a preliminary total radiant power.

The final total radiant power requires an additional correction for the relative distance to the Sun due to the eccentricity of the Earth's orbit[2]. The eccentricity, 0.0167, is the sine of the angle of maximum angular deviation from circular. It changes over a 100,000 year cycle with the relative position of major planets. It is near the minimum of its range.

In the northern hemisphere the orbit has its closest approach to the sun on January 2. Since this is near the winter solstice on December 22 the eccentricity cancels part of the effect of the orbit tilt in the northern hemisphere. The effect of the tilt is greater at most temperate latitudes and the maximum solar radiant power occurs when the path through the atmosphere is shortest, at noon of the summer solstice in the northern hemisphere.

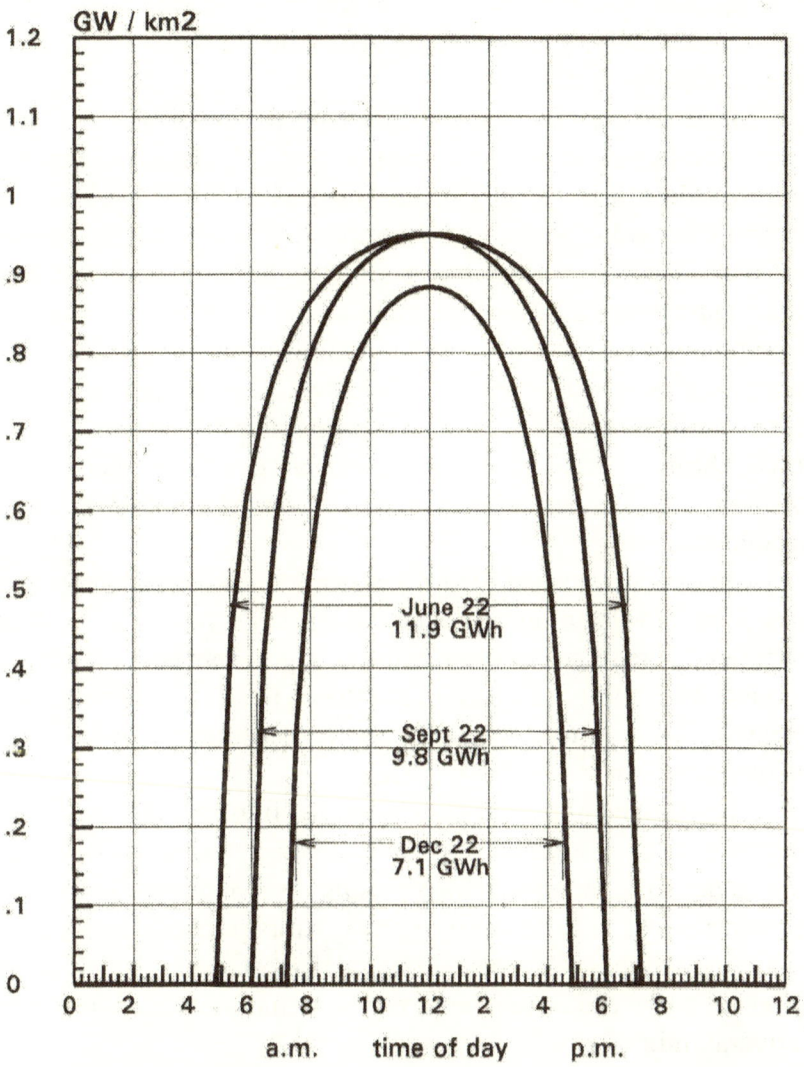

Fig. 10.3 Clear sky radiant power on the earth at 35 degrees north latitude

Fig. 10.4 compares solar power with the demand. This example shows the relative clear sky power at 35° north latitude on an equinox. The base of the solar power is raised to the level of the minimum daily demand to emphasize the similarity of the solar power and the peak demand.

Since solar power has no fuel cost the price of the power is particularly sensitive to the capital cost. Maximum utilization minimizes the effective capital cost. Three other power sources must contribute for solar power to be an effective source for the power grid. A base power source must supply nighttime power. Another source must cover the mismatch due to early evening demand. A back-up power source must provide reliability during periods when solar power is unavailable due to weather. The combination must meet reliability standards as well as maximum utilization of each source. The ultimate criterion is minimum price of power. Temporary transition performance criteria might hasten alternative resource development.

Agreement between the peak power demand and solar power gives solar power an added value. Using solar power as peaking power would create competition in the U.S. grid between solar power of doubtful reliability and natural gas that is flexible and reliable but expensive. The ultimate long term trend should increase the value for solar power in relation to natural gas.

The demand and solar supply curves illustrate the maximum utilization of both solar and base power. Each supplies about half the total demand. Increasing the spinning reserve of base power and creating a small reserve of solar power storage could provide the short term reliability. Underutilized high cost back-up power for periods of bad weather remains a problem that must be resolved by the overall mix of power sources.

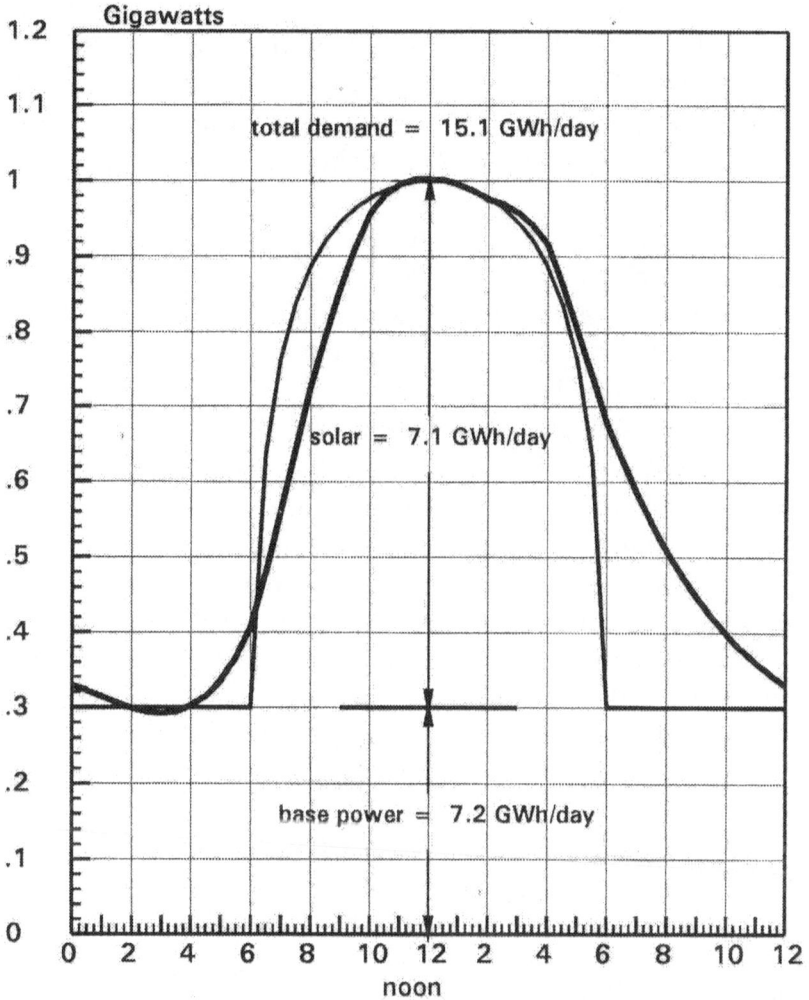

Fig. 10.4 Solar power at 35 degree latitude
compared with demand at an equinox

Note: Solar diplaced upward for comparison

Fig. 10.5 compares the seasonal solar power at various latitudes with the average relative demand. Each point on the solar power curves is the maximum total daily solar energy under a clear sky as shown in Fig. 10.3. The numerical value of the energy in kWh/day-m² corresponds roughly to the number of hours of exposure. The seasonal variation in solar power at all latitudes correlates with the length of the day.

The peak solar heating on June 22 precedes the peak demand for power by about 2 months. The specific heat of the Earth causes the maximum summer temperature to lag the maximum summer heating. The increased summer demand is consistent with use of air conditioning during high temperature.

The minimum solar heating on December 22 roughly coincides with the smaller peak winter demand. There is an absence of a significant lag. The winter demand correlates more closely with the greater demand for lighting than with temperature. The inverse correlation indicates that some other source must supply excess power in winter for solar power to be useful for the grid.

The minimum daily sunlight is zero at 66.55° north latitude on the winter solstice. At higher latitudes the winter period without direct sunlight increases. For an equal period at the summer solstice the sunlight is continuous. However the sun at high latitude is always at a low angle.

The annual solar energy variation is significant at all latitudes, but the variations are smallest at latitudes less than about 20°. About 40° is the maximum latitude at which solar power is economical for year round power. Interruptions by adverse weather are a problem at all latitudes. A degree of reliability is possible only under desert conditions.

The seasonal conditions produce a band of latitudes that accumulate snow. The area of snow accumulation acts as a seasonal heat sink that gives up heat during the winter and absorbs heat during the summer. The area is greater on the wider land surface of northern hemisphere. The depth of the snow accumulation is a heat sink that stabilizes the climate against longer term changes.

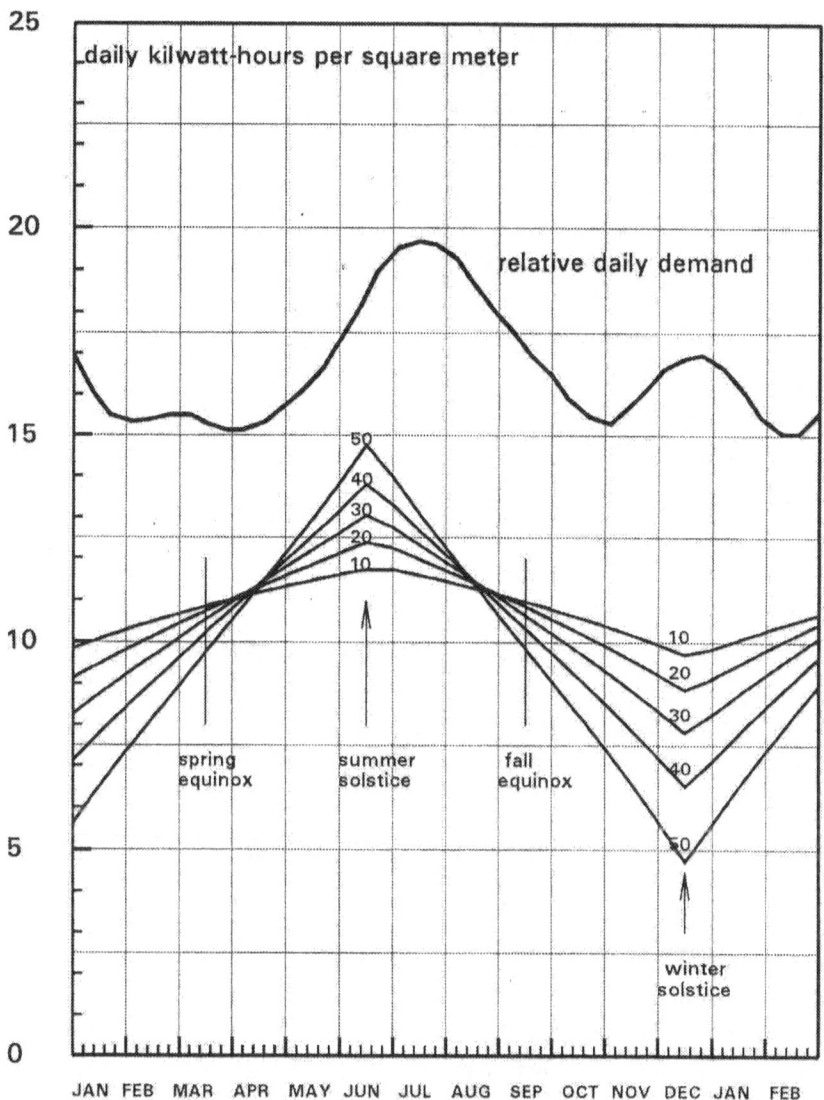

Fig. 10.5 Daily total solar energy at various latitudes compared with relative seasonal demand

197

Solar Photovoltaic Power

Photovoltaic power excites electrons in a semiconductor to conduct a current. The electrons in semiconductors have bound states with distinct energy separated by a *band gap* from Fermi states which allow electrons to flow freely as in metals. Photons that produce a conduction electron must have energy that exceeds the band gap. The electric power is the band gap energy in electron volts times the photoelectron current in amperes. The band gap voltage of a silicon photodiode is 1.12 electron-volts.

Solar-to-electric efficiency of solar panels has several origins. One is caused by the relation between the solar spectrum and the threshold voltage for exciting electrons. The other is the probability that electrons will recombine with the substrate without reaching the anode.

The quantum yield limit is the probability that a photon will cause an electron transition. This is associated with the spectral frequencies of sunlight. At spectral frequencies below the band gap energy the quantum yield is zero and the energy becomes waste heat.

Fig. 10.6 shows the response of a silicon photodiode to clear sky solar radiant power. The shaded area is the maximum power solar photoelectrons can produce. At a spectral frequency equal to or greater than the band gap energy the quantum yield depends on frequency but is less than one. The energy of these photons in excess of the band gap energy becomes waste heat. Only photons that exceed the band gap energy can approach the maximum efficiency of one.

The clear area that represents the energy of photons below the band gap appears as waste heat. The clear area above the shaded area is the energy of photons in excess of the band gap voltage. This is also waste heat. Band gap voltages over a broad maximum between 1.0 and 1.2 eV use about 50% of the radiant energy. The average quantum yield over all frequencies above the band gap energy limits the theoretical spectral efficiency of silicon photodiodes to about 44%.

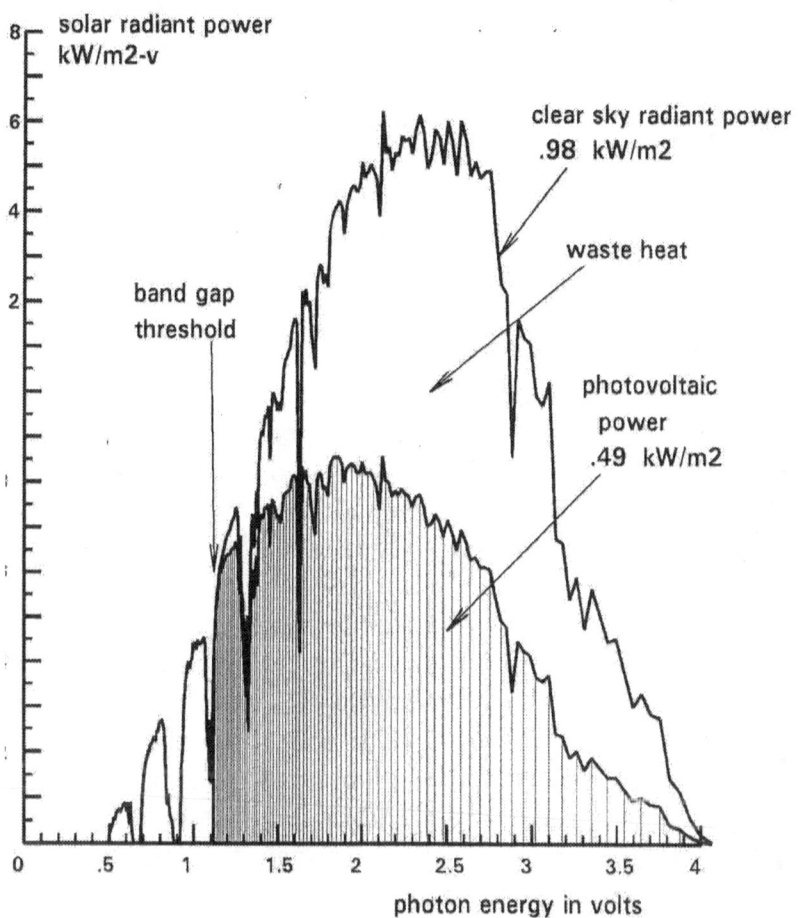

Fig. 10.6 Photovoltaic power from sunlight

Data source: NASA Technical Report R351

Recombination loss results from transitions of electrons in the free conduction state back to a bound state. These electrons are excited to the free conduction states by solar photons but fail to reach an electrode. The recombination probability depends on the electron density in the semiconductor. The density increases with the square of the current. It also depends on the length of the path through the semiconductor.

A junction potential loss occurs at the interface between the semiconductor and each electrode. It reduces the apparent external voltage due to the change in thermodynamic free energy of electrons at an interface between dissimilar materials. Junction potentials are a property of all interfaces between dissimilar materials.

An Ohmic resistance loss results when the flow of electrons in a semiconductor create an internal voltage that subtracts from the external voltage and power. This voltage is proportional to the square of the electron current and to the length of the electron path through the semiconductor. The path length is essentially the thickness of the semiconductor wafer.

Siemens Solar advertises silicon semiconductor panels doped with copper-indium-di-selenide that convert an average of 11.4% of the incident solar power to electrical power. This compares with an industry average of less than 9%. The conversion efficiency does not directly affect the cost. It is not the deciding factor in whether photovoltaic power can make a useful contribution to the grid. Understanding the low efficiency is important to applications that require a high level of power with minimum mass, such as power in space.

Photovoltaic technology is a variant of the technology used to produce other semiconductor devices[3]. Molten liquid silicon solidifies as a high purity multi-crystalline ingot. Parallel wire saws slice the ingot into wafers 0.3 mm thick. Treating the sides with p-type and n-type dopants and heating the wafer in a diffusion oven forms a p-n junction diode with a large area, typically 10 cm square. Vapor deposition or ion sputtering in a vacuum applies a metal electrode to each side of the wafer.

A mask covering everything but the electrode grid leaves one electrode surface transparent to sunlight. Vacuum deposition applies an anti-reflection coating of titanium oxide or silicon nitride to the front surface. A glass front cover and plastic back plate seal the assembly with electrical connections protruding from the edges. Interconnecting individual cells produce a panel with the desired voltage and power.

Photovoltaic panels respond to diffuse radiation with some reduction in power. This is an advantage over systems that require an optical focus of the sun to concentrate radiation to a useful temperature. Photovoltaic technology is useful over a less restricted range of locations than high temperature solar thermal power.

Photovoltaic power is suited to small scale applications. The rooftop area of most residences can provide the space for a major share of the power they use. It gives an economic value to land area that is otherwise unused. Tall buildings and higher density power require land area specifically devoted to solar energy collection.

An important direction for photovoltaic development is thin film technology that is less expensive, more efficient, and lighter weight. A base that supports the elaborate special treatments required for a monolithic integrated circuit is unnecessary. Producing a simple diode by the technology that creates central processors of computers seems like overkill. A film 100 nm thick is enough to have the required semiconductor properties. At the same time it reduces the thickness that restricts the efficiency. The electron path could be a factor of 30 shorter. The limit is an overall efficiency of about 33%.

Technology that stacks semiconductor films with different band gap energy in series could remove much of the spectral inefficiency. The overall output voltage of cells of a multi-layer stack connected in series is the sum of the band gap voltages. The upper layers of the stack must be transparent to the frequency of layers below them. The efficiency at each voltage balances the same electron current through each layer.

The cost of photovoltaic power is primarily the capital cost of semiconductor panels. At the present state of technology the panels are expensive. The opportunities for economies of scale are not large. This increases the attractiveness of their use as a distributed power source. Properly sealed cells have a 25-30 year operating life with little or no maintenance.

Table 10.1 is a self consistent summary of the typical economics. The calculation is for one kilowatt of installed nameplate power. The present cost of solar panels from about a dozen competitive suppliers is about $4500 per kW. This is roughly triple the capital cost of coal fired power. The absence of fuel cost and most operating and maintenance cost allows a higher capital cost. Most of the higher price of power is due to more limited utilization.

Table 10.1 Cost per kilowatt for a photovoltaic system

Effective Revenue	
Nameplate power	1.0 kW
Utilization	0.75
Effective price	$0.15 /kWh
Total electric power	2682 kWh/y
Total revenue	$401 /y
Operating cost	
Heat to electric efficiency	0.11
Maintenance	$20 /y
Capital cost	
Amortization period	25 y
Return on investment	0.08
Capital cost payment	$382
Allowable capital cost	$4780 /kW

The utilization factor is for a facility in a dry arid climate that receives 75% of the clear sky power at that site. The value used in the table is the 9.8 kilowatt-hours per day at 35° north latitude. The combined limit by hours of daylight, path through the atmosphere, and weather is typically about 30% of the maximum clear sky power. This is a more realistic utilization to compare with coal.

Solar Thermal Trough Power

Solar thermal power collectors concentrate radiation to produce a high temperature. The optical configuration of the mirrors distinguishes several systems. They depend generally on parabolic mirror surfaces that focus radiation to a smaller cross section by forming an image of the source. Whether the concentrated radiation is an actual image is irrelevant.[4]

A solar trough collector is a mirror whose cross section is parabolic in one dimension and linear in the other. The focus is an elongated image of the Sun. The heat receiver is a tube containing a heat transfer fluid that flows along the mirror focus. The heat ultimately drives a heat engine.

Rotating the collector axis at 15° per hour compensates for the Earth's rotation. At the 7.6° angle of the figure the trough points to the sun at 11:30 a.m. Adjustments for latitude are generally small in relation to the length of the trough.

The solar trough receiver is a tube that carries heat transport fluid at the focus of the trough. It couples the heat from solar radiation to a fluid such as Monsanto Therminol which carries the heat to a central heat exchanger. The heat exchanger heats steam that drives the steam turbine heat engine. Trough collector temperatures are marginally adequate for an efficient heat engine. At higher temperatures a satisfactory heat transfer fluid would become a problem.

Heat loss from convection and infrared thermal radiation are both potentially large at the operating temperature. The techniques that minimize them pay particular attention to the spectral properties of the receiver tubes in relation to the solar spectrum. Only a small fraction of the solar radiation spectrum overlaps the thermal emission spectrum at infrared wavelengths.

The coating of the receiver tube has high absorptivity and low reflectivity at solar wavelengths. The tube surface absorbs most of the solar energy. At infrared wavelengths where a black body emission is strong the tubes have high reflectivity and but low absorptivity and emissivity.

Enclosing the tubes in a vacuum jacket of special glass further reduces heat loss by eliminating thermal convection. The glass has selective optical properties. It is transparent at most solar wavelengths but reflective in the infrared. Sealing the vacuum jacketed tubes and lines carrying heat transport fluid maintains a permanent vacuum. The solar-to-thermal heat efficiency of the system is about 60%.

Fig. 10.7 is a ray diagram of a parabolic trough collector. The width of the Sun as viewed from Earth covers about 0.53°. Radiation from the Sun is a 0.53° ray bundle that converges to each point on Earth. The incident and reflected ray bundles at any point on the mirror converge from the edges of the Sun then diverge at the same angle following Snell's Law. The ratio of the area of the mirror to the area of its solar image is the *concentration ratio* in suns.

The mirror in Fig. 10.7 has a concentration ratio of 70 suns. The maximum clear sky radiant power at the image is 70 watts/m². A 1054 K *radiation temperature* means that a black body at 1054 K would emit the same radiant power. The temperature of the receiver tube is a steady state between radiation input, thermal radiation loss, and the temperature difference needed to transfer heat to the heat transport fluid. The fluid carries heat in proportion to the input-output temperature difference of the fluid.

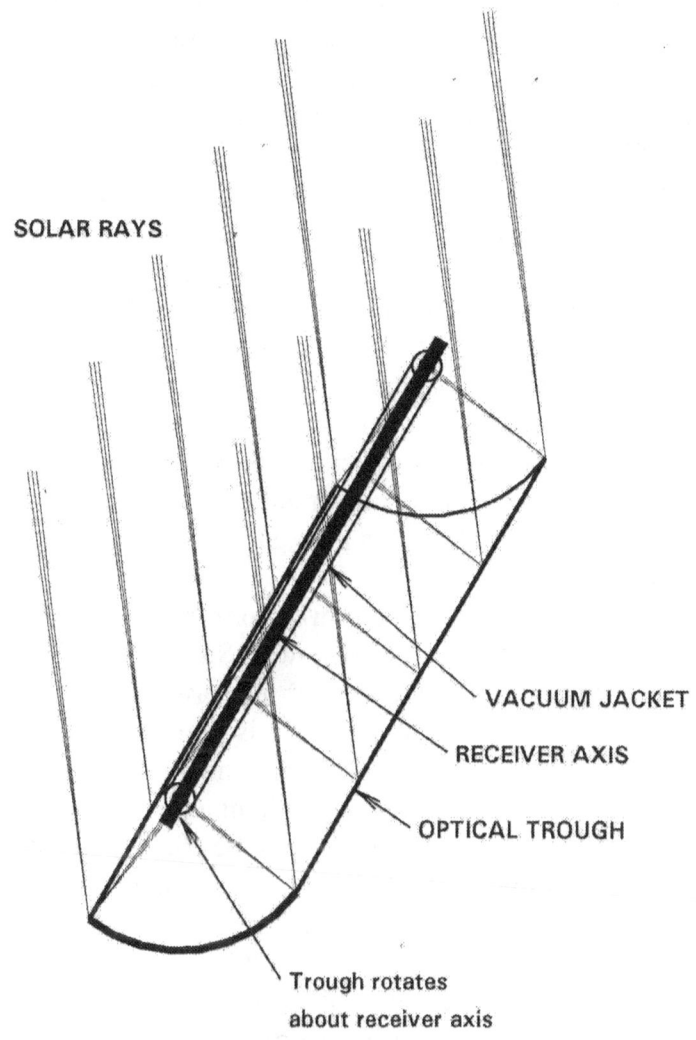

SOLAR RAYS

VACUUM JACKET

RECEIVER AXIS

OPTICAL TROUGH

Trough rotates
about receiver axis

Fig. 10.7 Solar parabolic trough optical elements

The Kramer Junction Company[5] operates the most significant commercial solar power installation in the United States in the Mojave Desert of California. Separate 35 MW units deliver a combined 350 megawatts of electrical power. A typical unit is an array of 800 trough mirror units each 162 feet long with 188,000 m^2 reflective cross section. The peak solar radiant power of 188 MW delivers 35 MW of electric power for a nominal overall solar-to-electric efficiency of 18.6 %.

A combination of solar power and natural gas heat the steam that generates the output power[6]. In the U.S. the Federal Energy Regulatory Commission rules allow a *qualifying facility* to combine renewable power with natural gas up to an average 25% of the total heat production and still qualify for favorable renewable power rates[7]. Each of the Kramer Junction plants can operate as a pure solar power facility, as a pure gas fired facility, or as a hybrid combination.

Maximizing the solar power component also increases the natural gas consumed for solar power. Since solar power matches the peak power demand this may decrease the total natural gas consumption. The operating criterion is to maximize production of the solar power component. This is generally consistent with the interest of the grid. Each kilowatt-hour of solar power used for peaking power reduces the quantity of peaking power that must be supplied by natural gas by 3 kWh. At the same time the 25% natural gas allowance counteracts most of the misfit and unreliability of solar power as a peaking power provider. The economic basis of solar power is the relative cost of solar power and natural gas.

The 2-stage steam turbines operate on the Rankine cycle. In order to produce high pressure, high temperature steam the heat exchange has two paths. One heats steam by solar power. The other heats steam with gas. The gas boiler produces higher maximum output temperature and pressure than the solar output. However, to achieve maximum revenue it is necessary to convert all solar energy input to electricity. The plant operator monitors the system temperatures by adding gas heat to optimize the total performance without exceeding the 25% limitation on gas power.

Table 10.2 shows an allocation of revenues and cost that is roughly consistent with the Kramer Junction operation. The value of electric power depends on when it is used. The historic demand is a basis for assigning rates such as 12, 10, 6, and 4 cents per kWh to each of the 8760 hours of the year[7].

Table 10.2 Combined cycle trough solar economics

Revenues

Rate Class	total hours	solar hours	electric GWh	$/year thousands
peak	732	732	24771	2973
near peak	4257	2675	60348	6035
off-peak	2799	61	688	41
far off-peak	972	0	0	0
total	8760	3468	85807	9049

Operation and maintenance	$ 2262 thousand
Capital cost	
Capital payment	$ 5972 thousand
Amortization period	25 years
Return on investment	.08
Allowable capital cost	$ 84.8 million

The first column is the number of hours of the year in each rate class. The second column is the number of hours of historic demand in each rate class that is typical for this latitude. Solar power completely covers the peak demand from noon to 6 p.m. of June through September. It covers the near peak power before late afternoon and evening and a few early morning off-peak hours but no far-off-peak hours. The Kramer Junction facility at 35° north latitude has a bias toward satisfying the peak demand hours. A typical plant has 188,000 m² of mirror surface. The historic solar power at this site averages 7.44 kWh/m²-day over the year.

Fig. 10.8 is a flow diagram of the combined cycle solar and natural gas plant operation. It shows the solar trough collector segments schematically at upper left. The heavy lines through the collector indicate the flow of heat transport liquid that collects the heat at the focus of the solar collectors. The closely spaced lighter parallel lines are flow lines that deliver steam. The lettered rectangles are heat exchange components that add, remove, or exchange heat. The circles are proportioning valves that control the flow of heat exchange fluid. The figure gives typical operating temperatures and pressures at key points in the cycle.

Table 10.3 lists the steps of the Rankine cycle that apply to the plant cycles in Fig. 10.8. In the figure the heat exchange component letters indicate the operation at each state of the cycle. The input to solar boiler **A** is the complete field of solar collectors that evaporate the liquid water, heat it to saturated steam, then superheat the steam. The input to solar boiler **B** is the gas heater. The gas heater produces a higher output temperature than the solar collector field.

Table 10.3 Rankine cycle for the power plant

1. **Adiabatic expansion to steam plus water**
   Solar super-heater **A** output, out, turbine-1 input
2. **Reheat to superheated steam**
   Turbine-1 output, gas heater C input, solar heater D input
3. **Adiabatic expansion to steam plus water**
   Gas heater, solar heater **C** output, turbine-2 input
4. **Condense steam to water**
   Turbine- 2 output, steam condenser **E** input
5. **Pressurize water**
   Steam condenser **E** output, pump input
6. **Heat water to saturation temperature**
   Pump output, solar boiler **A** input, gas boiler **B** input

In the maximum solar mode the proportioning valves maximize heat transfer fluid flow through heat exchanger **A**. In the stand-by solar mode the proportioning valves circulate heat transfer fluid flow through heat exchanger **D** to maintain a minimum standby temperature.

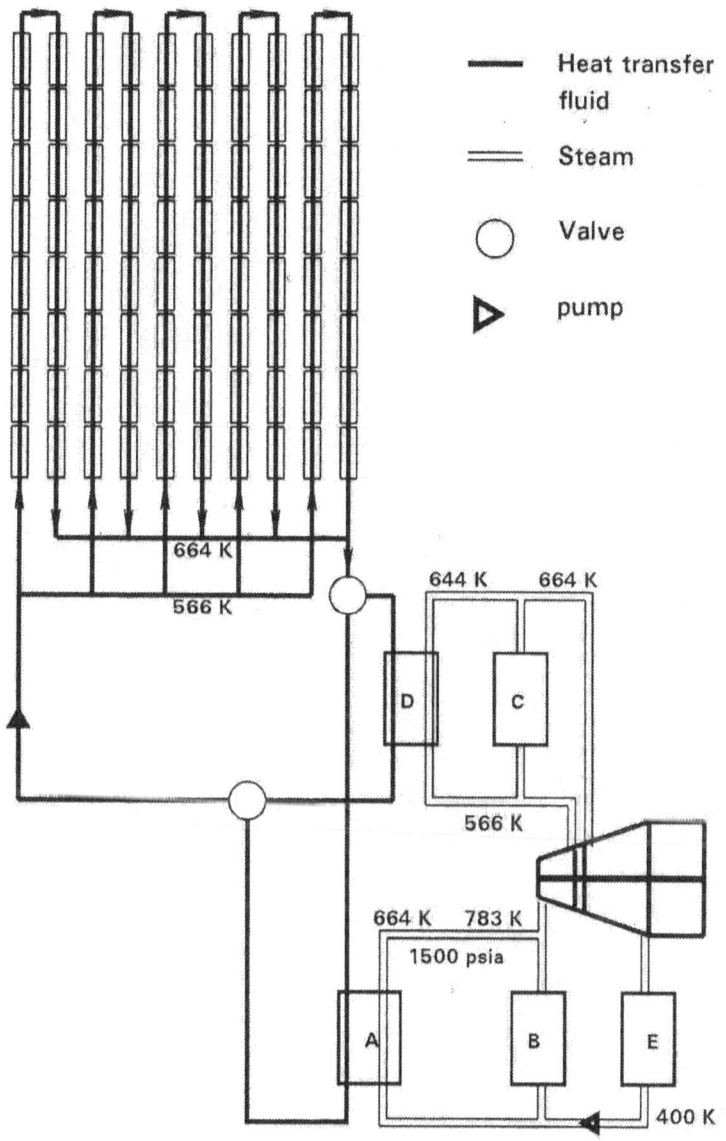

Fig. 10.8 Block Diagram of Kramer Junction System

The range of solar-to-heat conversion efficiency, from 55% in mid-summer to 20% in mid-winter, has at least three components. The increasing winter angle of the sun with respect to the fixed north-south axis of the receiver decreases the effective area of the mirrors.

The fraction of the heat lost in transporting it to the power generator depends more or less directly on the flow rate or residence time in the tubes and connections. The value of the natural gas component of the power is partly to provide backup and partly to increase the temperature and efficiency of the system. Any optional use of gas is to enhance the peak solar power in summer.

The heat-to-electric conversion efficiency including the engine efficiency ranges from 18% in summer to 6% in winter. The low efficiency in winter is characteristic of a system that is operating near the lower limits of its operating range. This is partly due to the smaller quantity of solar energy in winter. It is partly to take maximum advantage of the energy at the peak power period in summer.

Although combined cycle trough solar technology plants play only a minor role in supplying peaking power to the grid it is significant and could well grow larger. Solar power replaces almost four times the quantity of natural gas that is used to support it. In favorable locations combined cycle trough solar technology can in principle be the major peaking power supplier.

Dish-Engine Solar Thermal Power

A solar dish collector is a mirror surface with parabolic cross sections that are symmetric about the focal axis in two dimensions. Solar rays parallel to the axis form an image of the Sun at the focus of the parabola. Since the concentration is in two dimensions the nominal concentration ratio of the radiant power is equal to that of an equivalent parabolic trough squared.

The collector operates as a dish-receiver combination that tracks the Sun on two axes as a unit. Moving the large collector area favors a light weight membrane design with a reflective coating. The membrane stretches over the edges of a solid circular dish to form an air tight container. Reducing the pressure in the dish stretches the membrane to a rough parabola. Air pressure controls the focal length. The dish-engine-generator is a rigid assembly of circular dishes supported by 2-axis gimbals. The weight-to-strength ratio limits the size. The maximum dish diameter produced by present technology is 17 meters in diameter. An array of 3 dishes can collect 680 kW_h of solar power.

The assembly rotates about the longitude axis at the 15° per hour rate of the Earth's rotation. Rotation about the latitude axis ranges from near zero at each solstice to ±23.45° per day at each equinox.

A Stirling engine has a free floating piston that circulates the engine working fluid from one end of the cylinder to the other in a closed cycle. Action coordinated with the reciprocating motion of the piston mimics the Carnot cycle. The piston opens paths that allow the fluid to circulate. It shuttles alternate conductive and insulating contacts between the gas and the collector focus. It also alternates conductive and insulating contacts between the gas and the waste heat sink. The heat flow efficiency through the alternating thermal contacts limits the actual efficiency to about 23%.

The output of the largest present dish Stirling solar power systems is 150 kW_e of electrical power. Like other solar power systems the role of dish solar power in supplying power to the grid is to provide peaking power.

Each gigawatt of base power requires peaking power equivalent to 1200 dish-engine solar units. Although a dish-engine power system delivers power during periods of high demand, it has no inherent energy storage to provide near-peak power during the early evening or back-up to provide reliability.

Tracking the north-south angle of the sun by rotation about the latitude axis gives the dish-engine system relatively constant solar-to-electrical efficiency over the seasons of the year. The effective north-south cross section of the mirrors is essentially constant. The close coupling between the solar heat at the focus and the engine working fluid prevents a large increase in heat loss due to the reduced heat flow during winter. In this respect dish-engine power is superior to pure trough solar power.

Table 10.4 shows the economics of dish-engine power based on the prices used for solar trough power. The calculation applies a constant 23% solar-to-electric efficiency to the hours allotted to each rate class.

Revenues at the same rate as the combination trough-solar system ignore the absence of back-up power to provide reliability. At the $3500/kWh capital cost these revenues would allow more than double the capital cost of coal fired power. Determining whether these cost projections are realistic requires further operating experience and development.

Table 10.4 Limits of 150 kW dish-engine solar economics

Rate class	total hours	solar hours	electric kWh	$/year
Peak demand	732	732	114485	13738
Near peak	4257	2675	418370	41837
Off peak	2799	61	9540	572
Far off peak	972	0	0	0
Total	8760	3468	543395	56148

Operation and maintenance	$ 14,036 /yr
Annual capital cost payment	$ 42,110 /yr
Return on investment	0.08
Amortization period	25 yr
Allowable capital cost	$526 thousand

Solar Central Receiver Power

A flat mirror reflects essentially the same radiant power as the Sun itself, 1 sun. Superimposing radiation from 1000 flat mirrors on the same cross section produces a 1000 sun nominal radiant power neglecting the additional losses to be described. *Solar Two* is a 10 MW$_e$ central receiver on the Mohave Desert of California that is near the largest practical scale[8].

The collector is an array of 1916 flat heliostat mirrors. Each 42 m^2 mirror reflects solar radiation to a receiver at the top of a 300 foot central tower. The radiant power of the superimposed reflection is 600 suns. This corresponds to a radiation temperature of 1938 K. The solar cross section of each mirror decreases with increasing sun-mirror-receiver angle according to the time of day, time of year, and position of the mirror in the array. The area of the reflected radiation increases as it diverges from the mirror at the 0.542° angle due to the finite width of the Sun. Mounting the receiver on a central tower minimizes the angles and distances of the superimposed reflections that concentrate the radiation.

The receiver at the top of the central tower is a cylindrical array of vertical tubes 6 meters high and 8 meters in outer diameter. Although the surface coating of the tubes has high absorptivity for solar radiation and low emissivity for infrared thermal radiation they lose about half the input radiant power by radiation and convection. The radiation loss is obvious from the bright glow of the receiver. Note that high absorptivity means high emissivity and low reflectivity. All surfaces tend to be black bodies at high temperature.

A heat storage system uses the heat of fusion of a partially molten nitrate salt. Partially molten salt flows through the receiver tubes as a heat transfer fluid. It delivers molten salt to a storage tank at the bottom of the receiver. The heat of fusion stores heat for about 4 hours beyond the period of useful sunlight. In the absence of interruptions by weather Solar Two can deliver power on demand as a pure solar facility without fossil fuel back-up. The low viscosity and thermal conductivity of the salt are limitations.

Solar Two is a development and proof of principle project that cost $140 million to build plus $1.8 million/year to operate and maintain. The melting temperature of the salt sets the operating temperature of the engine. An engine temperature of 850 K is too low for efficient gas turbine operation but high enough for maximum steam turbine efficiency. Table 10.5 summarizes the characteristics.

Table 10.5 Solar Two characteristics

Site area of site	95 acres (2million sq ft)
Tower height	98.4 m (300 ft)
Collector mirrors	1916
Reflective surface area	81,350 m²
Concentration ratio	600 suns
Maximum heat collected	39 MW
Engine input temperature	849 K
Engine output temperature	561 K
Maximum electrical power	11.1 MW
Solar/thermal efficiency	0.48
Thermal/electrical efficiency	0.28

Table 10.6 summarizes the economic constraints on cost of a central receiver power system. The heat storage system stores the off-peak solar energy as heat and delivers the electric power during the period of near-peak electric power demand.

Table 10.6 Solar central receiver economics

Rate class	total hours	solar hours	electric GWh	$/year thousands
Peak demand	732	732	8.0	962
Near peak	4257	2736	30.0	2997
Off peak	2799	61	0	0
Far off peak	972	0	0	0
Total	8760	3468	38.0	3959

Annual operation and maintenance	$ 989 thousand
Annual capital cost payment	$2,969 thousand
Return on investment	0.08
Amortization period	25 years
Initial capital investment	$37 million

High Temperature Central Receiver

An alternative central receiver design could produce power in 1MW units more comparable to the largest wind turbines. Wind power and solar power are both low density energy sources that occupy large areas of land per unit of power. Unlike wind power, solar power on Earth blocks other use of the land.

High efficiency solar thermal power would take advantage of the high temperature of the Sun. The Sun becomes a high temperature source by concentrating the radiation. The dish-solar system achieves high concentration with a single optical surface that becomes increasingly difficult to align with the Sun as the size increases. The central receiver system achieves high concentration using a large number of mirrors. The concentration ratio of a central receiver can be increased by using mirrors with a slight curvature. This reduces the receiver input aperture area. The size of each collector mirror image changes as the sun-collector-receiver angle changes during the day and year. The superimposed reflections have a jumbled variety of sizes and shapes[9]. The extremes during the year fall within about ¾ of the area of a mirror. The receiver aperture area has a concentration ratio of 2000 suns.

The radiation provides a working fluid temperature high enough to take advantage of the Brayton cycle turbine engine efficiency. It is more compact than other solar power systems and gives more power per acre of land. Each of the 1080 mirrors is 2 meters square. The array reflects up to 4 MW of radiant power to the central receiver. Balancing the height of the back of the mirror array with the height of the receiver tower lowers the profile of the system.

To minimize the cost of the individual mirror units light weight polystyrene mirrors are cast as thin rigid plates. The vapor coated aluminum surface has a silicon oxide overcoat that permits occasional cleaning. The mirrors are mounted on gimbals driven by small, low cost stepping motors in a configuration that allows the same computer signal to drive all of the mirrors.

Fig. 10.9 shows the collector mirror array. Each 4 m^2 mirror reflects solar radiation to the aperture of a receiver cavity and concentrates the radiation cross section from 200 cm to 150 cm in width. The figure illustrates an array at a northern hemisphere location. The optical axis of the array is on the local north-south axis facing south. The kinematics of the mirror mounts allow independent correction for the Earth's rotation (east-west tilt) and seasonal and latitude adjustment (north-south tilt).

The scale of an individual module has an upper limit due to the size of the structures and a lower limit due to temperature loss associated with the surface to volume ratio. At a scale less than about 1 MW the insulation for this system either allows inordinate heat loss or requires an inordinate share of the heat. Above 4 MW insulation is not a critical issue. The large number of mirrors allows each mirror assembly to have minimum mass. Mirror adjustments require a minimum number of lightweight components using the smallest, least expensive stepping motors.

Computer driven stepping motors adjust each axis of each mirror in the array. An adjustment of 0.01° is typically a 2 cm movement of the 150 cm wide image. The adjustment mechanism kinematics allows each mirror to use the same stepping motor signal to track the Sun. At 0.01° per pulse the time between pulses is long enough for each unit to receive the signal in sequence. This averages the maximum power of the mirror and decreases the power ratings of the driving signal equipment by nearly a factor of 1000.

The receiver is a *black body cavity* that receives radiation though an aperture at the focus of the collector[10]. The radiation temperature at the receiver aperture is 2437 K. The high radiation temperature exists only at the opening of the cavity. Inside the cavity the radiation spreads to cover a much broader area of the walls at a lower radiation temperature. The heat loss by re-radiation through the input aperture is 332 kW or 8.3% of the input radiation. A door closes the input aperture whenever the solar input falls below the thermal radiation output.

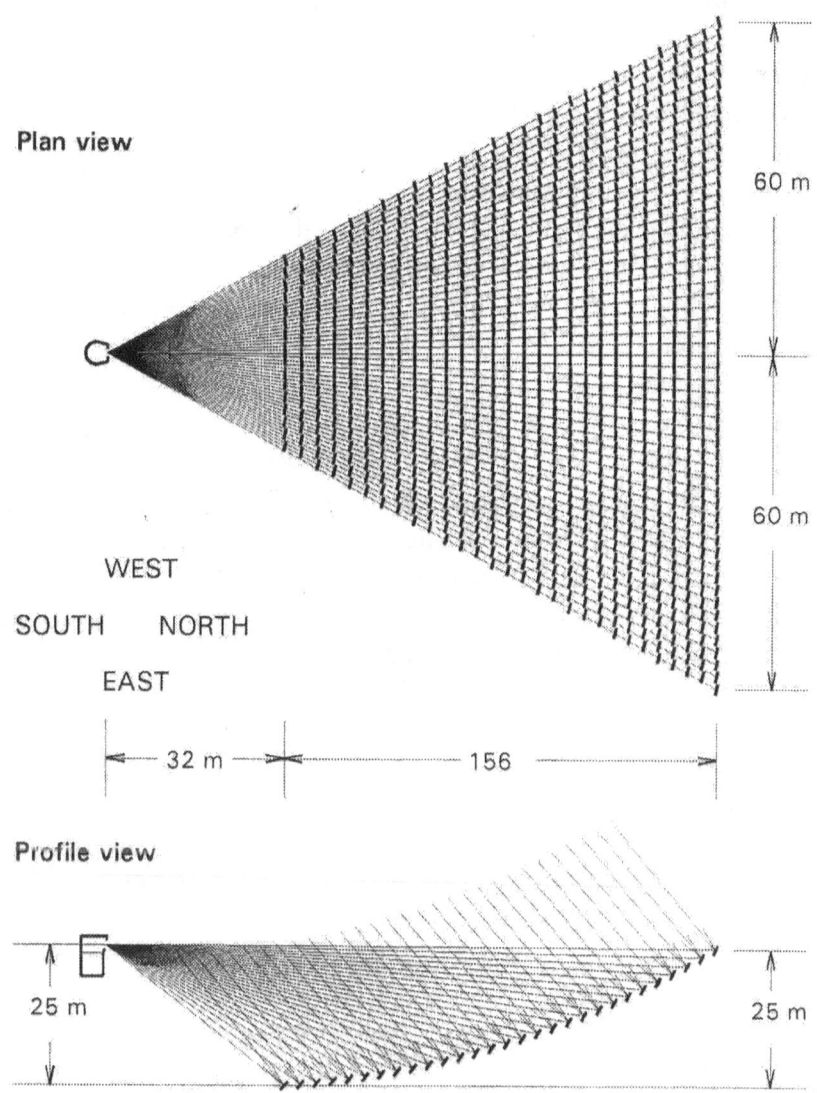

Plan view

60 m

60 m

WEST

SOUTH   NORTH

EAST

32 m   156

Profile view

25 m

25 m

Fig. 10.9 High temperature central receiver sysytem

Mirror angles at noon 22 March for a site at 40 degrees latitude

The flow rate of the heat transport fluid controls the dynamic equilibrium inside the cavity by continuously removing heat to compensate for incoming radiation. Two types of heat transfer distinguish the inside walls of the cavity. The incoming radiation falls only on surfaces of tubes that carry heat transfer fluid[11]. The balance between radiation input and heat transfer to the heat transfer fluid establishes a first approximation to the operating temperature. A temperature of 1100 K is within limits of available materials. Heat flow by thermal radiation is large at this temperature. All of the inside walls redistribute heat by thermal radiation and lose heat by thermal conductivity.

The heat transfer fluid consists of iron balls sealed in an argon atmosphere. High thermal conductivity gives metals desirable heat transfer properties. The high specific heat of iron allows it to transport heat efficiently. An auger can pump iron balls as a fluid. Concentric tubes with a counter flow of high temperature balls in the center and low temperature return on the outside can transport heat to the base of the receiver with minimum loss.

Table 10.7 summarizes the attributes of a module with a nameplate electric power capacity of 1.25 MW. The heat engine characteristics are for a gas turbine operating in a closed Brayton cycle using argon working fluid.

Table 10.7 High temperature central receiver characteristics

Area of site	20,640 m^2
Tower height	25 m
Mirror scaffold height	25 m
Collector mirrors	1080
Reflective surface area	4320 m^2
Concentration ratio	2000 suns
Engine input temperature	1120 K
Engine output temperature	540 K
Solar-heat efficiency	0.85
Heat-to-electric efficiency	0.45

Table 10.8 summarizes the economics of the high temperature central receiver module. The capital cost consists of collector mirrors, the receiver/heat storage system, the heat engine/generator, and the site preparation and structures.

Table 10.8 High temperature central receiver economics

Revenues

Total solar hours	13872 /y
Solar-to-electric efficiency	0.38
Total electric power	5271 MWh / y
Price	$0.10 / kWh
Total revenue	$527 thousand
Operation and maintenance cost	$ 105 thousand

Capital cost

Annual capital payment	$ 421 thousand
Return on investment	0 .08
Amortization period	25 years
Capital investment	$ 5.2 million

A fully automated system could operate with only occasional maintenance or adjustment. However even this requires more specialized knowledge, skill, and organization than $105,000 per year can provide. It is obviously more realistic to operate and maintain 665 modules generating 3500 GWh/y on a budget of $69 million/yr. A system of modules must be combined to achieve the economies of scale that take full advantage of the high temperature central receiver.

A thousand modules that deliver the equivalent of a conventional power plant require 4 km^2 of land. This is less than an acre/MW compared with the 9 acres/MW used for Solar Two.

Energy storage is necessary to provide both reliability and timing that matches the timing of solar power production with the timing of demand. One choice is to integrate energy storage with solar power generation by storing it as latent heat of fusion of a salt. Stored heat of fusion requires 8,000-12,000 tons per GWh of storage material depending on the substance and fraction that must remain molten.

Lava such as that from the Kilauea Volcano in Hawaii has the correct melting temperature. It can supply the necessary heat and tonnage at low cost. It requires about 30 tons/GWh.

One possibility is to design the lines that transport heat between modules to a central generator as a concentric heat exchanger. The central tube would contain high temperature iron ball heat transport fluid. This would be surrounded by heat storage material that would melt in layers adjacent to the heat transfer line. The temperature of the outer layers would decrease to the fluid return temperature. There they would provide insulation for the fluid return lines.

Notes and references for chapter 10

1. The effect of the curvature of the earth is neglected since its effect is at large angles where the radiant power decreases to zero. The relative distance through the atmosphere is then determined by the sum of a north-south component due to latitude and tilt and an east-west component due to the Earth's rotation. Donald Rapp, *Solar Energy*, Prentice Hall, Inc. 1981 Chapter 2 *Solar Geometry* derives the angle of the Sun as a function of the hour of the day, length of the day, and day of the year.

2. The solar radiant power outside the atmosphere at the nearest approach of the Sun on January 2, the perihelion, is about 7% greater than at the aphelion on July. 4. Below the atmosphere the tilt of the Earth's axis and the eccentricity of the orbit have the opposite effect in the Northern hemisphere.

3. Photovoltaic technology and cost are reviewed by H.A. Ossenbrink in Chapter 2 of *The Future for Renewable Energy 2: Prospects and Directions*, European Commission, UREC Agency, James and James, London, 2002

4. Non-imaging optics was defined by W.E. Welford and R. Winston, *Optics of Nonimaging Concentrators*, John Wiley and Sons, New York, 1978. A variety of reflective surfaces with specifications different from those required to produce an image can concentrate radiation. Compound parabolic concentrator reflectors can produce radiation temperatures limited only by the temperature of the source.

5. The Kramer Junction Company is the operating company formed when the Luz Corporation reorganized as a company that produces solar trough components, Solel Solar of Israel, and separate operating companies in different parts of the world. It has now provided over 15 years of year-by-year engineering improvements and improved operating practices.

6. Scott D. Frier, Gilbert E. Cohen, and Robert G. Cable, A*n Overview of the Kramer Junction Parabolic Trough Solar Electric Systems*, Kramer Junction Company.

7. The Kramer Junction plants are *qualified facilities* for a Public Utility Regular Power Purchase Agreements for emerging technology under Federal Energy Regulatory Commission. The agreements regulate time-of-use provisions for hybrid use of fossil fuel.

8. Solar Two was erected at a site in the Mojave Desert near Kramer Junction by a consortium of aerospace engineering firms. It was sponsored by the Sandia National Laboratory and the National Renewable Energy Laboratory. A similar installation at Sandia Albuquerque is used to simulate radiation from nuclear explosions.

9. The off-axis angle decreases the focal distance from f_0 to f in proportion to the cosine of the angle of incidence, $f = f_0 \cos a$. The receiver aperture is the maximum width of the radiation from any mirror at any angle during the year. The width of the radiation at the focal distance of a mirror with width, w_0, and focal length, f_0, is $w = w_0 + (f_0 - f) \tan \delta_{sun}$, where $\delta_{sun} = 0.542°$ is the angle subtended by the Sun from the Earth.

10. The emissivity and absorptivity of solids increases as the temperature increases. The reflectivity and transmissivity decrease. Surfaces inside a cavity at high temperature reach a common *black body temperature* as radiation from hotter surfaces is absorbed by cooler surfaces. Objects observed in a furnace while it is heating disappear in a uniform glow as the furnace reaches incandescence. Glass becomes opaque as it approaches the melting temperature.

11. The radiant power at the 2.25 m² input aperture is 2000 kW/m². As the radiation spreads the radiant power decreases to 100 kW/m² at surfaces which absorb it. Therefore 100 kW/m² of heat must be removed from 10 m² of active surface. The inactive surfaces inside the cavity equilibrate by the flow of thermal radiant power from higher temperature to lower temperature surfaces in proportion to the 4th power of the temperature ratio according to the Stefan-Boltzmann law.

Chapter 11
Imagining the Future

The picture that emerges from the preceding discussion is a nearly certain inexorable decline in the dependability of the carbon based fuels that provide most of the energy for electric power. Radical change is not necessarily imminent. What is certain is that the voracious demand for electric power guarantees that each decision to add new power to the grid requires increasingly complicated choices among resources and conditions. No future decision about new electric power will be based on the same facts, or possibly even on the same technology, as the one it supplements or replaces. In some combination of size, accessibility, energy, and environmental quality the cost of fossil fuel is almost certainly in an irreversible rise.

Nuclear power can only delay the inevitable. The small fraction of the fissile isotope in natural uranium limits the present enriched uranium supply. Plutonium enriched uranium could increase the supply by nearly a factor of 100. Nuclear weapon security and radioactive contamination problems complicate each increase in reliance on nuclear power, whether it is from isotope enriched uranium or plutonium enriched uranium.

Each individual share of the world's natural resources grows noticeably smaller. As the population approaches 10 billion humans adopt more diverse modes of individual productivity. More of the individual productivity necessary to survive and flourish now relies on energy in general and electric power in particular. With the world's largest populations only beginning to recognize this, there is more reason for a significant increase in demand than a decrease if that is physically possible.

Recognized alternative energy sources have limited ability to provide power in the required quantities with reliability and low cost. It is an open question whether distributed combinations can add up to a satisfactory answer.

The one exception to this generally gloomy outlook is the solar radiant power outside the Earth's atmosphere. The world's foreseeable energy requirements could be met by the solar radiant power through about 0.2% of the area available in a geostationary orbit band 38,500 km above the equator. The radiant power is constant and reliable for 24 hours a day, 365 days a year. This is at least 8 times more energy than is typically available in the same area on the Earth's surface.

Over a quarter century ago NASA commissioned a consortium of engineering firms to study the feasibility of space solar power as the primary base power source for the grid. An outline of the plan appeared in an article by Peter Glaser.[1] A Hearing of the Subcommittee on Space and Aeronautics of the 106th Congress in 2000 reaffirmed the conclusion that necessary technical concepts are well established but reported no significant progress.[2] Solar Power Satellites have remained a NASA program area, but its major project funds are committed to the manned space laboratory. What a demonstration project could usefully show has not been clearly defined. The reasons are a mixture of technology and economics.

The economic principle of space solar power is the same as any electric power. Current revenue must pay for the cost of operation, maintenance, and return the capital with interest. The revenue produced by a power plant is the nameplate power multiplied by the sale price and plant utilization factor. New applications, such as transportation, and time-of-use pricing should allow 100% utilization of electric power capacity by the time space solar power could be fully implemented.

A 100-megawatt pilot plant module operating continuously at the nameplate power would generate 876 GWh of energy per year. At the 6 cents per kWh average wholesale price of power from coal annual revenue would be $52.5 million. Compared with the $375 million price NASA sets for each military space cargo delivery using the shuttle the cost of transportation appears to make space solar power economically out of the question.

The economics and technology depend on scale. The needs of an inevitable population of at least 10 billion people could easily be 5000 GW. This could yield $2.6 trillion in annual revenues at a low average price for power. A 100 MW space power pilot plant would be huge in comparison with typical current use of power in space. It is only a tenth of a typical power plant and smaller than the possible world demand by a factor of 50,000.

The essential conclusion is that low cost reliable power from space requires thinking in terms consistent with the size of worldwide electric power systems. This dwarfs the size of the present space programs. Over the next century or so the power system which evolved over the last century must be replaced and enlarged. This requires rethinking that is quantitative and long term regardless of what energy sources are used.

The fundamental economic question the technology must answer is whether the electric energy a space power system provides over its useful lifetime significantly exceeds the energy to build and operate the system. The technology must maximize each of the following quantities to succeed.

- Total energy delivered by a power plant over its useful life divided by the total mass of the plant
- Mass of the plant divided by the total mass lifted to orbit.
- Orbital energy of plant mass divided by the total fuel potential energy at lift-off
- The total mass of the plant must be a minimum
- The useful life of a plant must be a maximum

The mass of the plant goes through three stages in its transformation. The plant materials are produced under normal atmospheric conditions in the gravitational field at the Earth's surface. They must initially have a form that is compact enough for a space vehicle to transport through acceleration at least triple the Earth's gravitational force. Those forms must unfold into a plant with the minimum mass to function for an indefinite lifetime at zero pressure and zero gravity.

Space Solar Power Principle

Fig. 11.1 is a plan view of a space solar power plant. It remains stationary with respect to locations on the surface by circling the Earth synchronized with its 24-hour period at a point above the equator. The common axis of the collector and transmitting antenna is perpendicular to the Earth-Sun axis. The collector and antenna act as independent bodies whose orientation is maintained primarily by their great momentum. Thrusters periodically correct the residual drift.

The collector continuously faces the Sun. The larger area corresponds to the 14% efficiency of the available photovoltaic technology. The smaller area represents the upper limit of the efficiency that could be expected from further development of collector technology. In either case the collector surface is an array of low voltage elements connected in series to provide power at about 20 kilovolts for microwave transmission.

The transmitting antenna continuously faces the Earth by daily rotation with respect to the collector. To produce the microwave beam a large array of small klystron microwave generators is distributed across the surface of the antenna in what is known as a phased array. This allows the microwaves to be transmitted with little loss to a well defined area on the Earth's surface as a beam of *coherent plane wave radiation*. Radiation with the same frequency and phase throughout is also a characteristic of lasers and masers. The cross section of a beam of coherent radiation is a *diffraction pattern*. It has a strong central peak surrounded by circular rings of negligible intensity. The diameters of antennas that transmit and receive coherent plane radiation are proportional to the wavelength of the radiation and to the distance between the antennas.

A phased array follows the *Huygens' Principle* according to which waves are propagated as if each point on every wave front acts as the source of a new wave. Conversely if an array of separate sources have the same frequency and phase in some reference plane they produce a beam of coherent radiation perpendicular to that plane.

226

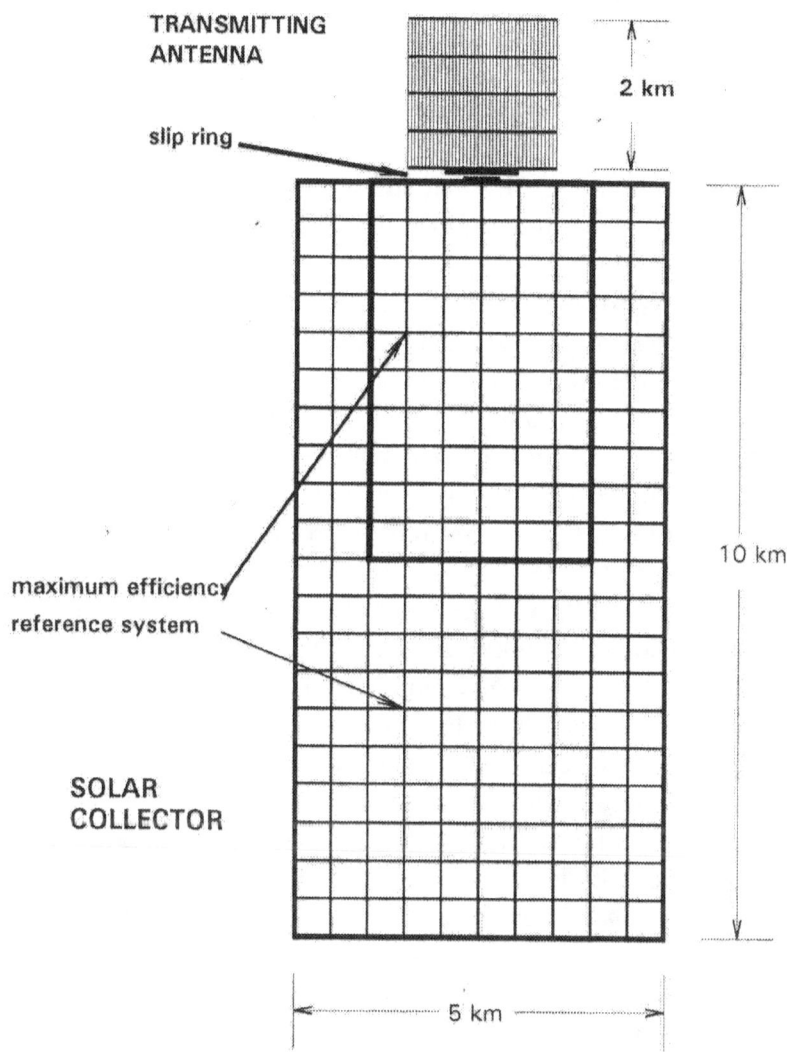

Fig. 11.1. Schematic space solar power system.

The vertical axis is in the same plane as the Earth's rotational axis with the collector facing the sun, The antenna continuously rotates to face the receiver.

A pilot beam from the center of the receiving antenna synchronizes the individual generators of the array. This aims them at the ground antenna. Each generator detects the pilot beam at a point in the same plane as its output. Each generator then acts as a microwave power amplifier that produces radiation with the same frequency and phase as the pilot beam in the opposite direction. It is not necessary for the outputs to originate in the same plane. This gives some latitude in the angle, planarity, and rigidity of the antenna.

The receiver antenna on Earth is an array of dipole antennas called a *rectenna*. Each element collects power in a relatively small area and rectifies it to a dc voltage. The frequency and phase of the incoming beam are then irrelevant and the planarity of the surface uncritical. Connecting the small voltages at each element in series gives the voltage transmitted to the power grid.

The dimensions of the receiver antenna are *proportional* to the wavelength and the distance between the antennas but *inversely proportional* to the corresponding dimensions of the transmitting antenna. The wavelength and distance determine the antenna diameters. Antennas of equal diameter would be 3.27 km in diameter. The 2 km transmitter diameter in Fig. 11.1 corresponds to a receiver diameter of 5.35 km. At locations which are not on the equator directly below the transmitter the cross section of the receiver is elliptical with a north-south dimension inversely proportional to the cosine of the latitude of the site.

A frequency in the industrial band at 2.45 GHz, a wavelength of 12.25 cm, is safely below water absorption that interferes with transmission through the atmosphere and is the principal biological hazard[4]. To take economic advantage of the size of the antenna requires 5 GW plant. At this power density the radiation temperature is about ¼ the normal sunlight, below body temperature and non-hazardous. This is still more than double the power the same area a photovoltaic surface delivers. The antenna can be transparent to sunlight. The land could have dual use as a greenhouse, for example.

A slip ring between the collector and the transmitting antenna provides a rotary electrical connection between them like the brushes on a dc motor. This major concentration of power must carry 430,000 amperes distributed in 20 kV circuits. Distributing the current around a large diameter ring accomplishes this. The inner ring is connected firmly to the transmitting antenna. The outer ring is secured to the collector through more flexible connections. The opposing rotation of the rings is driven by a stepping motor with a rotational period of 24 hours. This compensates for the frictional force of the electrical contacts around the slip ring to minimize their effect on independent momentum of the antenna and collector.

Ion propulsion pods are mounted at strategic locations on the collector and the antenna[3]. In the vacuum of space they provide thrust to correct the momentum that orients the collector toward the Sun and antenna toward the Earth. Each pod includes 2-axis stepping motors with communication to operators on Earth who control them. In the construction phase additional pods provide thrust to transport modules between near Earth orbit and the geostationary orbit. A 100 kW pod would deliver 4 tons to a geostationary orbit in about 12 days.

The ion source is a hollow cathode electric discharge through a gas. Positive ions drawn from the cathode cavity are focused at an opening in a 20-kv anode. They are ejected at high velocity. Although the ions use nitrogen or argon gas for momentum, the quantity is greatly reduced by the 20-kv acceleration as compared with the 15-20 volt energy of combustion reactions. The cathode is subject to erosion and requires occasional replacement.

The modular character of a space solar power plant is inherent in the limited mass that can be stowed in the cargo bay of a lift vehicle. A power plant requires many individual payload units. The total mass depends on the overall solar-to-electric grid power efficiency. Although the 1976 article by Glaser described a single large unit, it would be built as a collection of modules, packaged for transport to space, then unpackaged and assembled in space. The collector is constructed of modules 500 meters square. The antenna modules are 200 meters square. The modules would be constructed in near Earth orbit, transported to geostationary orbit, and assembled robotically.

Space Solar Power Feasibility

The absence of clamor for a crash government program is probably to the good. Space solar power has not yet crystallized as a commonly understood concept. The details must be described in somewhat the way a blind man would describe an elephant.

The system described in the Glaser paper is a baseline from which to start. A step-by-step analysis shows that beyond the relatively inefficient photovoltaic surface the loss in delivering the power to the grid is less than 50 percent[5]. The question is not the technical feasibility of the plant but the combined technical, economic, and political feasibility of constructing it in orbit.

The Manhattan Project to develop the atomic bomb that ended World War II is often accepted as the model for inventing and organizing systems to meet a challenging objective. The scientific and technical objectives were met when the bomb worked. Nuclear power is a bonus that happened to fit existing heat engine technology and economics.

The Apollo-Saturn Moon landing was a project to end the myth of technical superiority of the U.S.S.R. by showing that the U.S. could land men on the Moon, carry an artificial Earth environment in which they could function somewhat independently, and return to Earth. However the project had no end game. It simply ended, leaving a bloated though talented bureaucracy without credible long term objectives or funds to accomplish them.

The International Space Station used the experience of test pilots to establish a space flight capability resembling airline performance. It was successful in making the transition from ground to space reversible. The 30 years of subsequent experience made little progress toward manned space flight on a fast, reliable schedule.

Manned space programs show that astronauts can survive and drive a golf ball in the vacuum of space, but not very well. No construction worker can view these efforts as a way to accomplish major construction in space. Outside their engineering function of observing and testing flight characteristics, manned space missions show no compelling reason to place men in space.

230

Revenues that cover the cost are the ultimate test whether space solar power is feasible.. The revenues are known within a factor of two or so. A power station with a 5 GW nameplate capacity can produce up to $2.6 billion in annual revenue at the 6 cent per kWh price of present base power.

The size of the revenue can either be viewed as proof that space solar power is not feasible or as a challenge to see what would make it feasible. The financial target must guide the technology. The ideal system is a permanent asset with no operating cost. The costs must be capital costs that can be amortized over a very long term. These conditions act as a large multiplier of the allowable cost. They also impose technical challenges.

Ocean exploration, like space exploration, began with men in cumbersome suits probing no deeper than the upper few percent of the ocean depths. Serious useful work at true ocean depths requires robotic tools designed for that environment. Robotic video cameras permit better vision. Robotic tools can be operated from almost anywhere. Most importantly, robots are capital equipment that can function for decades on the power they help to generate. Humans in space are not only enormously more expensive. They are a short term operating cost.

Robots are simply tools, albeit sophisticated tools that would be required with or without human presence. Transforming bulk cargo into an operating power plant requires human direction, but that can be remote or pre-programmed. Changed thinking about robots must parallel changed thinking about structures.

Like hydroelectric power, space solar power could provide large revenues from low cost electricity. Low operating costs might ultimately justify the very large capital cost as well as sustain continuing capital projects. The projects involve large areas of land that is politically difficult to acquire in a growing population. The size of the facilities may be beyond the limit of venture capital to develop and build. Once built, the facilities are more or less permanent and may, like the Tennessee Valley Authority and Bonneville Power Authority, continue to operate as quasi-government agencies.

The energy to place mass in orbit is the sum of two components. The potential energy component is the energy needed to raise the mass to the desired elevation against the gravitational force. The kinetic energy component is the horizontal velocity needed to keep the mass in orbit. If the potential energy and kinetic energy are equal the orbit is circular. If the orbit parallels the equator with a 24 hour period it is geostationary above a point on the equator.

The function of the rocket is to accelerate the mass vertically through the Earth's gravitational field to an altitude where the horizontal kinetic energy at the peak of its trajectory prevents its subsequent fall from reaching the Earth. A near Earth orbit 200 km above the surface requires a speed of *mach 22.9,* that is, 22.9 times the 340 km/sec speed of sound at sea level. A high powered rifle accelerates a bullet to only about mach 3. To propel mass to space a rocket extends the acceleration over a longer distance by carrying propellant as part of the load.

Manned space flights limit the acceleration to 3g over a vertical distance of 45 km. Producing momentum in the lower part of the ascent decreases the fuel the rocket must carry to high altitude. The structural requirements for an unmanned vehicle might allow greater initial acceleration.

At launch the rocket consists of propellant, transport vehicles, and useful cargo. As propellant is burned its energy is distributed equally between forward momentum of the load and the reverse momentum of the burned propellant. Booster rocket vehicles separate and return to Earth when the propellant has been burned.

A rocket engine is essentially a nozzle shaped for maximum exchange of momentum. The shape of the nozzle parallels the shape of the output gas flow. This depends on external pressure. The rocket engine that propels a vehicle through atmospheric pressure is generally not the same engine used to adjust the trajectory, maneuver, and decelerate for the return from orbit.

The systems generally use low altitude booster rockets for the lift-off and initial acceleration. The cargo vehicle has a high altitude rocket engine for maneuvering and deceleration for return to Earth.

Table 11.1 calculates the energy change to go first to a near Earth orbit, then to a geostationary orbit. The gravitational acceleration and orbital velocity both decrease as the height of an orbit is increased. The geostationary orbital period matches the 24 hour day at an altitude of 35,860 km above the Earth.

Table 11.1 Properties of space solar power orbits[6]

Earth radius	6378 km
Gravitational acceleration	9.79 m/s2
Near earth orbit height	200 km
Gravitational acceleration	9.20 m/s2
Potential energy change	0.26 kWh/kg
Kinetic energy change	8.41 kWh/kg
Total energy change	8.67 kWh/kg
Orbital velocity	7.78 km/s (mach 22.9)
Orbital period	1.47 h
Geostationary orbit height	35860 km
Gravitational acceleration	0.22 m/s2
Potential energy change	7.10 kWh/kg
Kinetic energy change	-7.11 kWh/kg
Orbital velocity	3.07 km/s (mach 9.0)
Orbital period	24 h

Beyond near Earth orbit aerodynamic requirements disappear. A solar collector module can double as a solar powered space transportation vehicle after it is assembled in near Earth orbit. Added ion propulsion units can then propel it to geostationary orbit with additional cargo as needed. This eliminates a cargo vehicle and the energy and most of the propellant mass needed for transportation between orbits.

The rocket system efficiency is the ratio of the orbital energy of the payload to the fuel energy. Its value, roughly one eighth, is within sight of theoretical limits. It has three main components. Momentum is divided between the fuel and the vehicle. About half the original lift weight of propellant lifts unburned propellant. This depends somewhat on the time profile of the burn. The cargo accounts for about half the launch mass that is not fuel. This depends on the design of the entire launch package.

Fig. 11.2 shows the space solar power system greatly magnified in relation to the geostationary Earth orbit. The collector circles the Earth synchronized with a point on the equator. Rotational inertia keeps it facing the Sun. The collector and antenna are connected by the slip ring. Rotational momentum is added to keep it facing the Earth along a radial. In 3-dimensions the apparent eclipse period at midnight appears only at the equinoxes.

A pure cargo vehicle could improve the ratio of cargo-to-liftoff mass in several ways in comparison with the space shuttle used for flights to the international space station. The final landing mass is a bare bones orbiter without human support requirements. This decreases the mass to be decelerated and maneuvered for re-entry, the fuel to accomplish it, and the mass of the heat protection. Some increase in the fuel efficiency might come from a higher initial acceleration rate. A conservative target might be to deliver 100 metric tons of cargo into near Earth orbit using a system with the same launch mass as the space shuttle. This increases the mass of useful cargo by a factor of 4.

The collector mass is a major fraction of the cargo that must be placed in orbit. The schematic view in the Glaser article is unrealistic. It raises the question whether with no gravity or pressure to withstand a support structure is really necessary. A space solar collector is not so much a unified structure as a patchwork of stiff carpets. The magic property that makes it possible to convert rolls of cargo to stiff light weight surfaces and beams is the one natural resource space offers, continuous electric power in large quantity over endless time. The collector would be shipped to near Earth orbit as long rolls with photovoltaic coating on one side and connectors on the edges. Simple tools at the site would bend the edges to a shape stiff enough to maneuver and easy to connect to similar sheets.

The ultimate question that underlies any system design is whether the revenues can support the cost of transporting the necessary mass into geostationary orbit. The answer is in three further questions. What is the cost of transporting mass into orbit in relation to the value of the power? How much mass do the revenues allow? Is that mass sufficient to construct the plant?

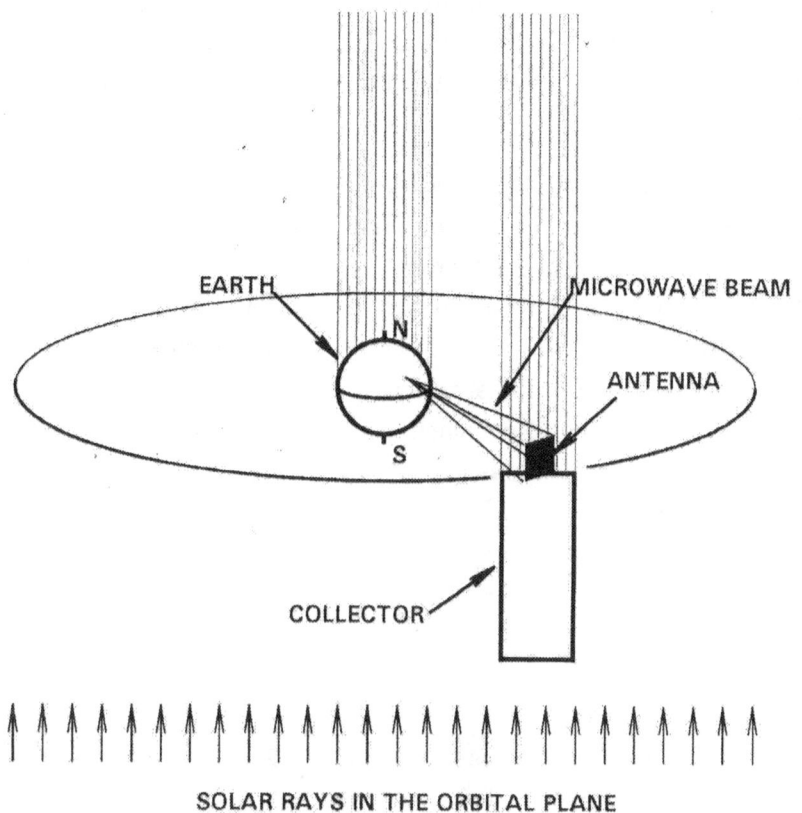

SOLAR RAYS IN THE ORBITAL PLANE

Fig. 11.2. Solar power from a geostationary orbit

The Earth and geostationary orbit have the same scale.
The space solar power system is enlarged X 4000.

Table 11.2 calculates the energy cost of mass in orbit. The calculation assumes that the rocket fuel is liquid hydrogen produced by electrolysis. The mass of the plant is then independent of the price of power.

Table 11.2 Fuel cost of transporting mass to orbit

Plant specific mass in relation to fuel energy

Energy of mass in near Earth orbit	8.67 kWh/kg-in orbit
Heat of combustion of hydrogen	242 kJ/mole
Fuel energy	3.73 kWh/kg-H2O
Fuel energy-to-energy orbital energy	0.125
Cargo-to-fuel mass ratio at launch	.053 kg-cargo/kg-fuel
Distribution of launch mass	
Payload mass	100 t
Vehicle mass	100 t
Fuel mass	1857 t
Total launch mass	2057 t
Fuel cost	
Electric power rate	$ .06 /kWh
Cost of hydrogen electrolysis	$ 2.02 /kg-H2
Other cost of hydrogen fuel production	$ 2.75
Total fuel cost	$ 4.75 /kg-H2
Fuel cost of mass in orbit	$ 86 /kg-payload

The calculation is based on the space shuttle performance assuming that the payload capacity increases to 100 metric tons on removing the requirement for a human environment. The vehicle that carries it to near Earth orbit has an equal mass. This makes the mass of the fuel and total launch mass approximately the same as the shuttle.

The fuel cost in dollars depends on the price of electric power that performs the electrolysis. The calculation uses the current price of power from coal, about 6 cents/kWh.

The next step is to calculate how much mass the revenues from the plant allow. The mass of an economically feasible plant is the capital allowable for rocket fuel divided by the cost of propellant per unit mass.

Table 11.3 estimates the limit of the mass that would make the plant economically feasible by allocating maximum revenue to transportation as a capital cost.

Table 11.3 Economics of space solar power

Revenues

Nameplate power	5 GW
Utilization factor	1.0
Average price	$ .06 /kWh
Annual power generation	43800 GWh
Annual revenue	$ 2628 million

Operating cost

Total operating cost	$ 600 million
Component replacement	$ 500 million
Robotic operations	$ 100 million

Capital cost

Amortization period	50 y
Return on investment	.06
Payment on capital	$ 2028 million
Plant replacement cost	$ 40560 million
Propellant (.3)	$ 12168 million
Vehicle operations (.4)	$ 16224 million
Plant components (.2)	$ 8112 million
Ground antenna (.1)	$ 4056 million
Fuel cost for mass in orbit	$ 86 kg
Allowable plant mass in orbit	141000 tons

The final step is to decide whether a 5 GW space power facility is compatible with a 141,000 ton mass. Like most numbers connected with electric power, the mass is too large to comprehend directly.

If the entire mass of a plant were represented by a sheet of stainless steel covering the area of the collector shown in Fig. 11.1 the sheet would have to be less than about 0.5 mm thick to meet the mass requirements. This obviously requires thinking that is different from simply transferring Earth bound engineering concepts into space. It requires a piece by piece re-examination of every component. This is necessary not only to minimize the mass but to maximize the performance in the absence of a human presence over a very long time.

Table 11.4 shows a set of relative masses that are consistent with the performance of current technology as given in the Glaser paper and total 141,000 tons. The performance data are largely retained in the 2003 report[7]. The 12% efficiency of photovoltaic technology is the only decisive improvement in efficiency that is possible. The mass of the components requires a fresh look at how the mass of every component can fit these specifications. An estimate of the mass of the plant and the cost of transporting it to orbit obviously requires assumptions about developments and designs that have not yet occurred. The assumptions that follow are only a basis for estimating an achievable mass.

Table 11.4 Mass limits for a 5 GW$_e$ system

	out / in	GW in	tons in orbit	flights
Power generation				
Solar power	0.93	75.40		
Collector	0.12	71.63	120938	1210
Power distribution	0.91	8.59	2833	29
Slip ring	0.96	7.82	1000	10
Microwave generators	0.82	7.50	12300	123
Transmitting antenna	0.93	6.15	3000	30
Receiving rectenna	0.87	5.72		
Power grid		5.00		
Total			140071	1400

The collector output from the photovoltaic surface totals 8.6 GW over 55 million m^2. Photovoltaic strips are connected in series to provide a standard plant voltage. About 20 kV is suitable for both the microwave generators and the ion propulsion position thrusters. Photovoltaic panels deposited on a 10 mil stainless steel base could satisfy the mass restrictions.

The divisions between the modules in Fig. 11.1 would be non-conducting I-beams that would be manufactured in space.

They would separate the 20 kV high and low voltage power buses as well as provide integrity for the collector structure. The mass of the collector is the mass of the stainless steel base. The structure that supports and isolates the electrical power could also stiffen the length of the collector.

Power distribution mass is based on 0000 gauge aluminum cable. The system is equivalent to 860 cables at 500 amps each over 5.5 km length. The weight is divided between the mass of the conductor and the isolation and support structures. New technology is needed to form insulation and structural supports in space. The tools would produce continuous lengths of I-bars and channels from rolls of low density composite materials and cure them for hardness strength and rigidity.

The slip ring design is limited by the need to separate the 20 kV conductors from the ground return lines in a 430,000 ampere concentration. The minimum diameter that can separate the 860 pairs of 500 amp contacts by 20 cm is 30 meters. At this diameter the slip ring would have a small fraction of the mass of the cables that connect it. It would also allow the slip ring assembly to be prefabricated on Earth in 3 or 4 sections.

Microwave generation requires 6.13 million 1 kW units clamped along the power bus structure of the antenna. Each unit, including the wave guide outlet, is allowed 2 kg. The vacuum of space should simplify the design.

The transmitting antenna is a lattice of 2000 I-beams manufactured in space from rolls of light weight composite material 2 km long extending from the slip ring connector. The top and bottom of the I-beams serve as a 20 kV power bus. The microwave generators are plugged on to the power bus at approximately 1 meter intervals giving roughly one generator per m^2. This distribution tolerates modest errors in the antenna angle.

The number of space flights needed to construct a single plant is the most daunting figure in relation to past experience. Previous space experience has consisted of a complete mission for each flight. A space solar power plant requires more than a thousand missions the size of the space shuttle. This requires new thinking.

The cost of a single plant cannot possibly support the infrastructure that is necessary to build it. This is where economy of scale comes into play. Dividing the cost of the infrastructure by 1000 space solar power plants changes the picture entirely.

The space solar power organization is a set of enterprises such as Table 11.5. They are ultimately all supported by the capital budget of the power plant operator. The major item for an individual plant is the propellant. This appears in the capital budget of the power plant operator but is a current operating cost of the space flight operator and the propellant manufacturer.

The design of a power plant that is essentially passive and can operate for the better part of a century on instructions from the ground appears technically feasible. The feasibility of an infrastructure to construct it and provide occasional maintenance and modernization depends on the economy of scale that can be achieved. The demand by the United States is probably marginal. The combined demand of North America, East Asia, India, Europe, and Russia is not.

The plant operation would monitor the performance, correct the orbit configuration with thrusters, and select modules for maintenance or replacement. The system would have enough redundancy to make isolated component failures irrelevant. The ground antenna operation is a part of ordinary grid distribution functions.

Table 11.5 Organization of space solar power enterprises
   Plant construction and maintenance infrastructure
       Plant component development and manufacture
       Space vehicle development and manufacture
       Propellant manufacture
       Launch facility maintenance and operation
       Space flight operations
       Space construction operations
       Plant maintenance operations
   Primary power generator functions
       Solar collector ion propulsion robotics
       Slip-ring connector ion propulsion robotics
       Transmitting antenna ion propulsion robotics
   Power distribution enterprises
       Receiving antenna operation
       Transmission line operators
       Independent service providers

The Alternatives

Daunting as space solar power may be there is no reason to assume that alternatives that are adequate are any less difficult. Regardless whether it is deployed the uncertainties in space solar technology deserve to be resolved to make it a quantitative reference for other options.

Hydroelectric power is by far the dominant renewable source. In the U.S. it supplies about 6% of the demand. It has about reached the limit of useful sites. Wind, solar, and biomass power are major alternatives. For any one of them to reach a level comparable to hydroelectric power would be a major achievement. In combination they would be 25% of adequate. Any talk of them being a part of the solution to future electric power, either separately or in some distributed combination, should be supported by some arithmetic.

Notes and references for Chapter 11

1. A summary of the report was published as an article in *Physics Today*. Peter E. Glaser, *Solar Power from Satellites*, Physics Today, February 1977

2. U.S. Congress, Hearing before the Committee on Science, House of Representatives, *Technical Feasibility of Space Solar Power*, September 7, 2000, Serial No. 106-87

3. Ion propulsion units developed for missions among the inner planets provide a 5-ampere ion current at 20 kilovolts. John E. Foster and Michael J. Patterson, *Microwave ECR Ion Thruster Development*, NASA Glenn Research Center, Cleveland, Ohio AIAA-2002-3837

4. See for example: C.K. Chou and A.W. Guy, *Immunological and hematological effects of microwave transmission from a satellite power system*, U.S. Environmental Protection Agency, Research Triangle Park, North Carolina

5. A popular account of the politics of space solar power and the options that have been considered for transporting cargo to orbit has been written by the aerospace engineer who led contributions by Boeing Aircraft Company and testified in the 2000 House of Representatives hearing, Ralph Nansen, *Sun Power: The Global Solution for the Coming Energy Crisis*, Ocean Press, Ocean Shores, WA

6. The potential energy of mass in orbit is measured as a decrease with respect to zero in outer space. The changes in Table 11.3 are the energy that must be added between the surface and near Earth orbit and between near Earth orbit and geostationary orbit.

7. Baseline data and efficiencies available from existing technology are given in a publication of the Congressional Board of the 97th Congress, *Solar Power Satellites*, Office of Technology Assessment, 2000

Chapter 12
Glossary

The glossary is an introduction to technology for the more serious non-technical reader. It may or may not be of interest to more general readers but deserves an explanation in either case.

Every technology requires a specialized vocabulary to express ideas precisely. Many terms are best defined by an equation that shows explicitly how terms are related. The syntax of technical language is usually the mathematics of problem solving. This is a continuing effort that never ends. The vocabulary by itself is an outline of the kind of problems a technology can solve.

The glossary is arranged by technology. The vocabulary is introduced in a natural order that expands the context of ideas as new words are introduced, the order the words would appear in a course of study to learn the syntax. Some words in the glossary do not appear in the text but are useful to complete the context of words that do appear.

Economics is a social science that deserves special mention as a technology. The social part is a subject of continuing research. The properties of the science part are useful to understand. Economics is a statistical science like thermodynamics. Statistical quantities have a distribution of possible values that depend on the size of the sample. Economics would resemble thermodynamics more closely if the number of samples were much larger. A billion billion billion molecules is a small thermodynamic system. Economics seldom deals with as many as a billion monetary transactions. The width of a distribution and predictability of the numbers is directly related to the size of the sample.

Prefixes to Measurement Units

Example: kilowatt = kW = 10³ watts = thousands of watts.

Deca	da	10	tens	deci	d	10^{-1}	tenths
Hector	h	10^2	hundreds	centi	c	10^{-2}	hundredths
kilo	k	10^3	thousand	milli	m	10^{-3}	thousandths
mega	M	10^6	millions	micro	µ	10^{-6}	millionths
giga	G	10^9	billions	nano	n	10^{-9}	billionths
tera	T	10^{12}		pico	p	10^{-12}	
peta	P	10^{15}		femto	f	10^{-15}	
exa E	10^{18}	atto a	10^{-18}				

Units of Measurement and Equivalents

Mass, kilogram, kg
1 lb = 0.4536 kg
1 t (metric ton) = 1000 kg
1 ton (short) = 907.1 kg

Length, meter, m
1 ft = 30.48 cm
1 in = 2.54 cm
1 mi = 1.609 km   1 Ws = 1 J

Area, square meters, m²
1 acre = 4047 m²
1 are = 100 m²
Force, Newton, N (kg m²/s²)
Pressure, Pascal, Pa   (N/m²)
1 Pa = 10^5 bar
1 atm = 1.013 bar

Volume, liter, l
1 m³ = 1000 l
1 qt = 1.101 l
1 bbl = 159 l (petroleum)

Power, watt, W
1 hp = 746 W

Energy, joule, J   (kg m²/s²)
1 btu = 1055 J
1 cal = 4.184 J
1 toe = 41.8 GJ (oil equivalent)
1 toe = 11.6 MWh
1 kWh = 3600 J
1 GWy = 8760 kWh

Economics

- cooperative systems that enable people to survive and prosper. Microeconomics is the effect of relative advantages in trade and commerce. Macroeconomics is the organization of national and international money and trade

Money is token representations of value issued by a national government for convenience in trade. It includes coins, bills, and other securities. Its use establishes its value. Business activity is limited by the amount of money in circulation.

The Central Bank is a bank established by a government to place its money in circulation by loaning it at low interest to commercial banks.

The Market is an abstract concept, not an institution. It is the sum of all transactions that involve an exchange of money. It includes exchanges between individuals, retail sales, wholesale sales, and other formal and informal institutions that act as brokers or agents that reconcile the cost to a seller and value to a buyer at a price that defines the value.

An efficient market is a hypothetical ideal market in which all of the factors that affect the sale price are equally transparent to the buyer and the seller, making the sale price as close as possible to a true value.

Productivity is the net income earned from producing or adding value to goods or from performing a service.

Inflation is an increase in the value of goods and services in relation to the face value of the money required to purchase them. Inflation occurs when the demand exceeds the supply. This may occur when there is more money in circulation than is justified by current business activity. In this case it is controlled by decreasing the money supply. Monetary policy usually promotes a dynamic economy by a low level of inflation that stimulates strong business activity with minimal changes in market prices over the short term.

Consumer price index is the ratio of the current market price of a representative sample of goods and services to the price at some reference time. It is used as a measure of inflation.

Monetary policy is the combination of government actions to maintain the value of its money by controlling the amount of money in circulation and by other measures to create public confidence in the value of money.

Capital is the money needed to start a business enterprise as opposed to money needed to operate it.

Capital account is the financial record associated with management and repayment of capital debt from current revenues.

Current account is the financial record associated with payments for the management, operation, and debt from current revenues of an enterprise.

Fiscal policy first sets tax rates that provide the revenues to operate government yet encourage the enterprise and commerce necessary for productivity. It then balances government spending for the infrastructure for productivity and public safety against entitlements to provide social equity in the face of unforeseeable dislocations.

Public investment is money spent by governments to create conditions which allow productive enterprise to flourish.

Private investment is money loaned by individuals or private organizations to create a productive enterprise. The loans are generally secured by bonds which repay the money at a specified interest rate or as shares of ownership in the enterprise.

National debt is a nation's capital account usually held as long term interest bearing bonds with a fixed date of maturity.

Capitalism is an economic system that enables private enterprises to start businesses that require equipment costing many times the expected annual revenues by proportioning the cost over a long period. It uses competition to produce goods and services at the highest quality and lowest price. Its disadvantage is the social dislocation when unproductive businesses are eliminated

Amortization period is the useful lifetime of equipment over which an initial capital cost is allocated and the capital debt repaid.

Communism – the failed economic system based on state ownership of productive property. The advantage in social stability was accompanied by the stagnation produced when productive enterprise becomes too complex for central control.

A stock market is a public market where shares in the ownership of privately owned corporations are bought and sold.

Price earnings ratio, p/e, is an indicator of the relation between the price and value of shares and the nature of the corporation. A *Value stock* of a mature company with dependable earnings might typically have a p/e = 15. A *Growth stock* of a newer company which has not yet reached its productive potential can have a much higher p/e ratio based more on its potential than its actual record.

Gross domestic product, GDP is the total earned income from the individual productivity, including foreign nationals, in a nation. It excludes goods and services produced in other countries.

Purchase price parity is a method of computing the GDP of underdeveloped nations to take account of local prices below the world market price.

A public corporation is a group of private individuals organized to produce goods and services that require more capital than individuals can provide. They can be either *for profit* or *not for profit*. Their capital is raised by selling shares of cooperative ownership on a market that is open to the public.

The global market results when trade between nations establishes a global price for goods, services, and labor.

Free trade is trade that allows all sellers to compete for buyers on the basis of price with no tariffs or artificial barriers that prevent the market price from reaching the level set by the most productive producers.

Balance of trade is the difference between a nations money spent on imports compared with the foreign money received for sale of exports.

Exchange rate is the value of a nation's currency in relation to that of another.

The foreign account is a nation's account of the balance between its money it spends abroad and the foreign money it receives in exchanges with foreign countries. Currency exchange markets correct unbalances in the overall accounts by changing the exchange rates.

The foreign current account is a nation's account of its money spent for imports and the foreign money received for exports, including exchanges during travel. In general, imports from another nation are bought with the importer's currency.

The foreign capital account is a nation's account of its money residents spend for investments in enterprises or property of a foreign nation and the amount of foreign money it receives for investments by foreign nations in its enterprises or property.

The balance of trade is the difference between the value of a nations money that is spent on imports compared with the total value of foreign money received for sale of exports.

Electrical power

- technical concepts of the electric power industry.

resistance - the component of a circuit load that dissipates power by doing useful work and generating heat.

reactance - the inductance, and capacitance components of a circuit load that absorb alternating current energy by shifting the phase of the voltage with respect to the current..

impedance - the sum of the resistance, inductance, and capacitance components of a circuit.

impedance matching - the power dissipated by the load in relation to the power dissipated in the circuit itself.

energy efficiency - ratio of the electrical energy delivered to the fuel energy consumed.

reactive energy storage - energy, proportional to the square of the current, stored by the reactance of a circuit.

nameplate power - maximum capacity of a generator.

rotating power - power currently being produced by a generator.

spinning reserves - nameplate minus rotating power, the reserves available.

pumped storage - a facility to store a reserve of hydroelectric energy for use at a later time by pumping water into a reservoir at a higher elevation.

chemical storage - a facility to store major quantities of electric energy for use at a later time by generating fuel electrochemically.

consumer classes - classification of electric power consumers by decreasing variability in demand as residential, commercial, industrial, or utility.

peak demand period - electric power demand during periods of maximum demand that defines the required maximum generating capacity.

utilization factor - the ratio of the power delivered by a particular power generating unit to its nameplate power averaged over a specified time period.

base power - power from facilities that produce a high volume of electric power at the lowest possible cost, but may have limited ability to vary the output.

peaking power - power from facilities with flexibility to change a high volume of electric power on hours notice at higher than minimal cost.

distributed power - power sources with varied capacity and timing distributed over many locations.

time-of-use pricing - contracts for time metered power for supply or delivery at prices based on the historical demand.

real time pricing - contracts for time metered supply or delivery at real time market prices.

avoided cost pricing - price premiums to smaller power producers based on the cost of new base power facilities that would otherwise have to be built.

local distribution center - a center responsible for dispatching power to the grid in a particular geographic area.

regional power pool - The U.S. is divided into three sets of interconnected local distribution centers which exchange power with adjacent regional power pools.

power dispatching priorities - rules that decide which of the available sources of power is next to be switched on or off line.

contractual obligation dispatching - The first consideration in dispatching power is to fulfill the contractual obligations to the supplier.

order of merit dispatching - electric power is dispatched in order of the sources that offer power at the lowest price in the absence of higher priority criteria.

out of order dispatching - dispatching to must-run generators or generators that would relieve grid congestion, regardless of market price.

power shedding - opening switches on isolated portions of the grid to relieve the congestion on a larger portion of the grid.

Thermodynamics

- relations among variables that describe all changes in temperature, pressure, volume, and chemical composition of equilibrium states of a system that conserve the mass and energy of the system plus its surroundings.

system - a quantity of matter distinct from its surroundings. In a closed system the mass of each chemical element is conserved, but more than one homogeneous gas, liquid, or solid phase and their solutions.

surroundings - the entities adjacent to a system that may exchange energy in the form of heat or work

phase - a homogeneous quantity of gas, liquid, or solid matter with distinct chemical composition.

state variables, Pressure, P, temperature, T, volume, V, and concentration - measurable properties of the state of a system. Internal energy, U, enthalpy, H, entropy, S, and free energy, F, - properties of the state of the system derived by thermodynamics.

equation of state, PV = nRT - ideal gas law, the relation among the state variables of an ideal gas. Thermodynamic properties of ideal gases are rigorous, by definition. Most gases are ideal at modest pressures.

path variables, δq, δw, increments of heat and work energy exchanged by a system and its surroundings to change the state of a system.

adiabatic process, $\delta q = 0$, a process that changes the state of the system without exchanging heat with the surroundings.

isothermal process, $dT = 0$, $\delta q \neq 0$, a process which exchanges just enough heat with the surroundings to maintain a constant temperature. The δ indicates an incremental exchange with the surroundings. The d indicates an incremental change in the state of the system.

solution - a homogeneous mixture of substances in the gas, liquid or solid phase.

solvent - the constituent of a solution in which other substances are dissolved. (The choice may be arbitrary)

solute - a constituent of a solution dissolved in a solvent.

molar concentration, moles of solute / liter of solution.

Molal concentration, moles of solute / 1000 grams of solvent. Molal concentrations give the mole ratio without the ambiguity of partial molar volumes.

Mole fraction, $X_a = n_a / n_a + n_b + \ldots$, the mole ratio of a substance in a solution of gases, liquids, or solids.

Partial pressure, $P_a = X_a P$, mole fraction of the total pressure of a gas.

Density, $d = g/cm^3 = kg/l = t/m^3 = Gt/km^3$

Stoichiometric coefficient, v_i, the number of moles of a substance, i, in the balanced equation for a chemical reaction.

Standard state - a reference state with standard system temperature, pressure, and concentration, usually T = 298 K, P = 1 bar, and concentration with a numerical value of 1 in whatever system is used. For pure liquids and solids X = 1, for solutions either X = 1 or M =1 molal, and for gases P = 1 bar. The ratio of the actual concentration to the standard state concentration keeps the numerical value but makes it a dimensionless ratio that can be used as an exponent or logarithm. Standard states may be fictitious. For example, water vapor cannot exist in its standard state at 1 bar and 298 K.

Internal energy function, $dU = \delta q + \delta w$, First Law of Thermodynamics, conservation of energy when a system absorbs external heat or work.

Entropy, $dS = \delta q_{rev} / T$, thermal definition for a reversible process in which heat is exchanged in a way that the temperature remains uniform and well defined.

Entropy, $dS > \delta q_{rev} / T$, thermal definition of irreversible processes, usually processes too rapid to maintain a uniform temperature and/or pressure.

Entropy, $S = R \ln W$, statistical definition. Entropy increases in relation to the number of ways in which the quantum states of a system are populated, a quantitative measure of randomness. A system has a large, but finite, number of states which increases with temperature. Since the number of molecules in a system is much larger than the number of states of a molecule, the fraction of molecules in each state is reproducible.

Entropy values - In the limit that a system has a single lowest energy state, the entropy of a system approaches zero at T=0 K. Entropy at a different temperature, pressure, and/or volume is calculated in steps using one of the following equations interspersed with the entropy of any phase changes.

$$\Delta S = C_v \ln(T_2/T_1) + nR \ln(V_2/V_1) = C_p \ln(T_2/T_1) - nR \ln(P_2/P_1)$$

Enthalpy function, $H = E + PV$, (definition), $dH - T\ dS + V\ dP$ (for an ideal gas), $\Delta H = C_p \ln T_2/T_1$ - for changes in temperature.

Enthalpy of combustion, $\Delta H_i = q\ /\ v_i$, The enthalpy of certain classes of reaction, including combustion, can be measured directly. If the reactants and products are in their standard states it is the standard enthalpy of combustion. The enthalpy under existing conditions can be converted to the standard conditions using the enthalpy equation given above.

Standard enthalpy of formation, $\Delta H_i^\circ = v_c \Delta H_c^\circ + v_d \Delta H_d^\circ - \Delta H_i^\circ$ the enthalpy change for the reaction that forms 1 mole of compound i from its elements in their standard states. The values ΔH_j° can be the enthalpy of any reaction of the same type, such as combustion.

Activity, $a_a = \gamma_a\ m_a$, $\gamma_a\ X_a$, concentrations modified by an activity coefficient, γ_a, to give equations for nonideal solutions with the same form as the rigorous equations for ideal gases.

Gibbs free energy function, $G = H - TS$ (by definition)

$dG = -\ S\ dT + nRT\ d\ln(P)$ - for an ideal gas

$\Delta G = nRT \ln(a)$ - for any substance at constant temperature

Reaction quotient, $Q = p_a^v\ p_b^v\ ..\ /\ r_c^v\ r_c^v\ ..$ a ratio of the activities of the reaction products, p_i, to reactants, r_j, with each p and r raised to its stoichiometric coefficient in a balanced chemical reaction.

Standard free energy of reaction - (reactants a and b; products c and d)

$$\Delta G^\circ = v_c \Delta G_c^\circ + v_d \Delta G_d^\circ - v_a \Delta G_a^\circ + v_b \Delta G_b^\circ$$

Equilibrium constant, $K_{eq} = Q_{eq}$, reaction quotient for a system at equilibrium where all reactants have reached their equilibrium activities.

Free energy of reaction, $\Delta G = \Delta G^\circ + RT \ln Q$

$\Delta G° = -RT \ln K_{eq}$ - chemical equilibrium at constant temperature. To proceed spontaneously the free energy of chemical process must decrease, $\Delta G < 0$. The condition for chemical equilibrium is $\Delta G = 0$.

Temperature dependence of equilibrium composition - (Clausius-Clapeyron equation) $\ln (K_2 / K_1) = -\Delta H° / R (1/T_2 - 1/T_1)$

Electrochemistry

- the thermodynamics of systems that exchange electrical work, E dQ, with the surroundings.

electrical work, $w = E\ dQ = E\ i\ dt$, the energy required to transfer an electrical charge, dQ coulombs (ampere-sec) to a system across a voltage difference, E.

electrochemical cell, a polar, 2-terminal device in which chemical reactions either cause an external current to flow or are caused by the voltage of an external circuit.

galvanic mode - an electrochemical cell in which spontaneous chemical reactions produce an external current and voltage.

electrolysis mode - an electrochemical cell using an external voltage and current source to induce a chemical reaction in the cell.

anode, the cell electrode to which anions flow and from which electrons flow to an external circuit. It is the positive electrode of the cell, the electrode at which oxidation, the loss of electrons, occurs. Example: $\frac{1}{2}\ H_2 - e^- \rightarrow H^+$

cathode, the electrode to which cations flow and to which electrons flow in an external circuit. It is the negative electrode of the cell, the electrode at which reduction, the gain of electrons, occurs. Example: $H^+ + e^- \rightarrow \frac{1}{2}\ H_2$

electrochemical equivalent, a unit equal to the molar concentration divided by the number of electrons it exchanges on oxidation or reduction.

Faraday constant, $F = 96{,}484$ coulombs/equivalent, the conversion factor converting concentration to units of electrical charge.

Nernst potential, $E_{rev} = -\ \Delta G\ /\ zF$, the voltage of a reversible electrochemical cell that corresponds to the change in free energy for the cell reaction, the cell reaction free energy in voltage units.

oxidation potential, the Nernst potential of a half cell reaction in the oxidation direction. Tables of standard oxidation potentials record the free energy of oxidation half cell reactions with the reactants in their standard state activities.

stoichiometric coefficient, z, the number of electrons that must be exchanged for a chemical reaction to proceed as written.

concentration potential, $E = -RT/zF \ln Q$, the galvanic cell voltage component

cell current, $i = z F \, dm/dt$, the amperes of current generated by a cell reaction that proceeds at a rate dm/dt equivalents per second. Positive current is the direction of the ion flow in the cell from anode to cathode.

overvoltage - the change in cell voltage that results from a finite current.

Radiant Energy Exchange

- the thermodynamics of energy exchange by electromagnetic radiation.

electromagnetic wave, the repetitive period over which the magnitude and direction of the electromagnetic field transverse to the direction of a beam of light completes a full cycle.

frequency, v sec-1, electromagnetic waves per second.

wavelength, λ cm, length of an electromagnetic wave.

wavenumber, ω cm^{-1}, number of electromagnetic waves per cm.

radiant power, P watts, power transmitted by electromagnetic radiation.

Note: To specify a limited spectral region the term is preceded by the word *spectral*, and its symbol is subscripted v, λ, or ω. The spectral limitation is incorporated in equations as dv, dλ, or dω.

radiant intensity, I watts/str, radiant power in a specified solid angle.

radiant energy, Q joules, radiant power over time.

black body, a surface that absorbs all incident radiation with no reflection or transmission through the absorbing body. Note that a body can only absorb, reflect, or transmit radiation. If there is not reflection or transmission it is a black body that can only absorb radiation. If it is a perfect absorber, conservation of energy makes it a perfect emitter.

transmissivity, $\tau = P/Po$, the fraction of incident radiant power transmitted through a surface.

reflectance, $\rho = P/Po$, the fraction of radiant power reflected from a surface.

absorptance, $\alpha = P/Po$, the fraction of radiant power absorbed at a surface.

Note: By conservation of energy, $\alpha + \rho + \tau = 1$.

Kirchhoff's law, $P = \varepsilon P_{bb}$, the thermal radiant power emitted by a surface relative to a black body surface at that temperature.

emissivity, ε, the material property defined by Kirchoff's law.

speed of light, $c = 2.99793 \times 10^8$ m/sec

Stefan-Boltzmann constant, σ_{SB} 5.670 x 10^{-8} W m^{-2} K^{-4}

Stefan-Boltzmann law, P = σ_{SB} A ε T^4, the radiant power from a surface with area A as a result of its absolute temperature.

work function, the energy, in volts, at which the kinetic energy of the electrons in a surface material barely exceeds their binding energy.

Planck constant, h = 6.6260 x 10^{-34} J-sec

Planck's Law, The spectral radiant power emitted by a black body surface. Most substances become black body absorbers at high temperatures. Planck's Law derives spectral radiant power of a free electron gas that obeys "particle in a box" quantum restrictions in absorbing and emitting radiation as independent oscillators with a continuous distribution of frequencies. At lower temperatures most substances have reflectance and/or transmittance, and thus emissivities less than one.

$P_{bb}(v) \, dv = 2\pi h v^3 / c^2 \, [e^{hv/kT} - 1] \, dv$

$P_{bb}(\lambda) \, d\lambda = 2\pi h c^2 / \lambda^5 \, [e^{hc/\lambda kT} - 1] \, d\lambda$

absorbance, A = -$\log_{10}$ [(Po-P)/Po]

absorption coefficient, k = A/b

absorption depth or path length, b cm, the length of the absorption path through an absorber.

molar absorptivity, a = A/bc

molar concentration, m, moles/liter of absorber

Physical Mechanics

    - the laws governing the motions of mass which conserve mass, energy, and momentum as they move in force fields or force free environments.

    Speed, velocity, $v = dx/dt$, m/s, rate of change in position. Speed is a scalar quantity, magnitude without regard to direction. Velocity is the vector component of speed in a particular direction.

    Acceleration, $a = dv/dt$, m/ s^2, the rate of change of speed or velocity of a body.

    Momentum, mv, kg-m/s, is the property of the motion of a body that is conserved (constant) in the absence of an external influence.

    Force, $f = ma = d(mv)/dt$, kg-m/s^2, the property of an external influence that changes the momentum of a body, causing it to accelerate or decelerate.

    Newton, N, unit of force with dimensions kg-m/ r^2.

    Newton's Law of gravitation – $F = G\, m_1 m_2/r^2$, the force of gravitational attraction between two masses separated by a distance, r, where G is the universal gravitational constant

    Gravitational constant, $G = 6.672 \times 10^{-11}\, m^3/kg\text{-}s^2$.

    Gravitational acceleration, $G = G\, m_1/r^2$, the acceleration of a mass m_2 on the surface of a much larger mass m_1 having a radius r. The mean value on the Earth is $g = 9.780$ m/s^2. It varies by ±0.275 % from the equator to the poles.

    Potential energy, $dE = f\, dx$, kg-m^2/s^2, the change in energy of a body due to a change in its position in a force field.

    Kinetic energy, $E = \frac{1}{2} mv^2$, kg-m^2/s^2, the energy of a body due to its speed.

    Joule, J, unit of energy with dimensions kg-m^2/s^2.

    Moment of inertia, $I = mr$, kg-m, where m is the mass of a body and r is its distance from an axis of rotation. Solid cylinder with radius r, $I = mr/2^{1/2}$. Solid sphere with radius r, $I = mr/2$. Collection of bodies, $I = \Sigma m_i\, r_i$.

    Frequency, f, cycles/s

    Angular frequency, $\omega = 2\pi f$, radians/s

Angular momentum, $I\omega$, kg-radians/s, is a property of a rotating body that is conserved in the absence of an external force.

Rotational energy, $E = \frac{1}{2} m\, r^2\, \omega^2 = \frac{1}{2}\, (I\omega)^2/I = \frac{1}{2}\, (2\pi rf)^2/m$, J, the rotational kinetic energy of a rotating system.

Harmonic oscillator, a vibrating system whose restoring force is directly proportional to displacements in any directions.

Force constant, $k = df/dx$, kg/s^2, the force per unit of displacement of a mass from its equilibrium position in a harmonic oscillator.

Vibrational frequency, $v = (k/\mu)^{1/2} / 2\pi$, s^{-1} for a harmonic oscillator system.

Vibrational energy, $E = \frac{1}{2} kx^2$, J, where x is the displacement from an equilibrium position,

Fluid mechanics

- describes the position, velocity, and acceleration of volume or mass elements of homogeneous fluids as they move in force fields or force free environments in ways that conserve mass, as they exchange energy and momentum with the surroundings.

Laminar flow has velocity stream lines parallel to the flow direction with a parabolic radial profile increasing from $v_x = 0$ at the wall to a maximum value in the center of the stream. It occurs at low flow velocities in channels with smooth surfaces. Pure laminar flow has no mixing between stream lines. Heat is transferred across the stream only by thermal conductivity.

Turbulent flow at velocities above a critical value has a flow cross section with a laminar sub-layer slowing to zero velocity at the walls, a transition layer that is neither laminar nor fully turbulent, and a fully turbulent central region which widens with flow velocity. The turbulent region has rapid radial mixing.

laminar layer thickness, δx, the boundary between the fast moving center of a turbulent flow stream and the slow moving edges. (See Nusselt number.)

incompressible flow, flow with fluid density independent of flow rate velocity at $v_x \ll$ mach 1.

flow cross section, A_f, the effective cross section area of a flow channel neglecting crevices inaccessible to flow.

flow diameter, $d_f = 2(A_f/\pi)^{1/2}$, the capacity of a flow channel expressed as the diameter of an equivalent cylindrical tube.

hydraulic diameter, d_h, 4 times the flow cross section divided by the wetted perimeter of the channel walls. It is the capacity of a channel in which the flow is limited by wall friction. Where d_f describes tubes with equivalent capacity to transport mass, d_h describes the diameter of a cylinders with equivalent capacity to transfer heat through the walls.

Examples: Cylindrical tube: $d_h = d_f = d$. Annular space between concentric cylinders: $d_h = d_2 - d_1$, $d_f = (d_2^2 - d_1^2)^{1/2}$. A number n parallel cylinders with diameter d: $d_h = d$, $d_f = n^{1/2}d$.

mass flow velocity, v_m, m³/s, mass per unit time passing a point in a continuous stream. It is independent of temperature, pressure, density, or cross section.

volume flow velocity, $v_v = v_m/\rho$, the capacity of a channel to transport fluid volume, depends on the local density of the fluid as well as v_m.

linear flow velocity, $v_x = v_v/A_f = v_m/\rho A_f$, the speed of mass flow channel averaged over the flow cross section.

flow kinetic energy, ½ $\rho V\, v_x^2$, flow kinetic energy of a volume of fluid.

flow power, ½ $\rho\, v_x^3\, A_f$, flow power of fluid passing through cross section, A_f.

hydrostatic pressure, the pressure exerted by a fluid at rest. It is the potential energy per unit volume of fluid. For flow in the direction of the pressure gradient, part of the potential energy in the flow direction becomes kinetic energy, ½ $\rho V\, v_x^2$.

stagnation pressure, the pressure registered by a closed probe (Pitot tube), facing directly into a flow stream, the potential plus kinetic energy per unit volume.

static pressure, $P_{static} = P_{stagnation}$ - ½ ρv_x^2 the pressure registered by a probe orthogonal to the flow stream.

head, p = $\rho g\, \Delta h$, Pa, the pressure imposed by a column of liquid with density, ρ, kg/m³, depth, Δh, m, and gravitational acceleration, g, m/s².

Bernoulli effect, $p_1\, h_1\, a_1$ - $p_2\, h_2\, a_2$ = $\rho V g\, \Delta h$ + ½ ρv_x^2, conservation of mass and energy for flow through a pipe at two points that differ in pressure, p_1 - p_2, height, h_1 - h_2, and cross section, a_1 - a_2.

viscosity coefficient, η, pa-sec, is the fluid property characterizing its mechanical resistance to flow. It is the tangential shearing force per unit of surface area per unit velocity gradient. (Note: 1 Pa = 1 N/m² = 1 kg/m-sec) The viscosity coefficient of a Newtonian fluid is independent of velocity. Long chain, non-isotropic molecules may form non-Newtonian liquids.

shear stress $\tau_z = \eta\, A_{yz}\, dv_z/dx$ is the force in the flow direction due to velocity gradient dv_z/dx between adjacent volume elements with area A_{yz} along the flow direction.

Poiseuille's formula, $v_v = \pi \, d_h^4 \, \Delta P \, / \, 128 \, \eta \, \Delta x$, relates linear flow velocity v_x in a cylindrical tube to the hydraulic diameter d_h and the viscosity η of a fluid. It is the integral of the shear stress over the radius of a tube with length Δx.

skin friction coefficient, c_f, an empirical function of viscosity. The Blasius relationship for fully turbulent flow through a cylindrical tube is $c_f = .079 \, Re^{-1/4}$.

viscous drag, $\Delta P_{visc} = \frac{1}{2} \, c_f \, \rho \, v_x^2 \, A_h / A_f$, the pressure decrease that occurs when a fluid moves through a channel at velocity v_x. It is flow kinetic energy converted to heat by friction at the wall.

Heat Transfer by Fluids

- the branch of fluid dynamics that describes predictable macroscopic exchanges during fluid flow that is chaotic on a smaller scale.

Buckingham pi-theorem - In test situations too complex for a rigorous description measured values of the variables can be expressed as an empirical function of dimensionless ratios of relevant variables. Well chosen ratios produce empirical laws that are valid beyond the conditions of the test case. Certain ratios are named for the individuals who demonstrated their utility. The following four ratios are particularly useful for problems of heat exchange by fluids.

Reynolds number, $Re = \rho \, v_x \, d_h \, / \, \eta$, increases with increasing flow rate and thus with increasing turbulence. It is a ratio of factors which favor turbulence to viscous forces which tend to prevent it. The transition from laminar to turbulent flow is in the range 2000 to 10000 for gases.

Nusselt number, $Nu = d_h \, h_c \, / \, \kappa$, relates heat transfer and thermal conductivity to the hydraulic diameter of the apparatus. The effective thickness of the laminar sublayer at the walls can be defined $\delta x = \kappa / h_c$. The Nusselt number is thus the ratio of the hydraulic diameter to the laminar layer thickness, $Nu = d_h / \delta x$.

Prandtl number, $Pr = \eta \, c_p \, / \, \kappa$ is the ratio of the viscous transmissivity to diffusive conductivity, a ratio that distinguishes the heat transfer capability of different fluid substances. It is independent of the flow conditions. It is large for viscous liquids, can approach zero for high conductivity liquid metals, and has a value near $Pr \approx 1.0$ for most gases..

Grashof number, $Gr = g \rho^2 \beta x^3 \Delta T / \eta^2$, ratio of buoyant forces to viscous forces describing the similarity of gases that conduct heat by free convection. The thermal coefficient of expansion of the fluid is $\beta = d\rho/dT$.

heat exchange area, A_h, the area of the walls of a given length of flow channel that exchanges heat between a fluid and the wall by conductivity through the flow boundary at the wall.

thermal conductivity coefficient κ (W/m-K), a property of solids and stationary fluids defined by the Fourier heat theorem.

Fourier heat theorem, $q = - \kappa A_h \, dT/dx$, the rate of heat flow through a temperature gradient dT/dx normal to heat exchange area A_h.

forced convection, $q = h_c A_h (T_{wall} - T_{gas})$, heat exchanged with the wall of a tube in which gas flow is induced by a blower or pump.

convective heat transfer coefficient, $h_c = Nu \, \kappa/d_h$ where

$Nu = 0.023 \, Pr^{0.4} \, Re^{0.8}$ for turbulent flow at $Re > 10,000$

$Nu = 0.052 \, Pr^{0.4} \, Re^{0.3}$ for laminar flow at $Re < 2,100$

The value of Re adjusts the effect of differences in the apparatus. The dependence on Pr accounts for differences among gases. Physically h_c is the thermal conductivity of the laminar sub-layer, $h_c = \kappa/\delta x$.

free convection, heat exchanged with a surface at which gas flow is induced by flow thermal density gradients.

buoyancy, $\tau_z = g \, \Delta\rho = g \, b \, \Delta T$, is the force per unit volume in the vertical z-direction due to gravitational acceleration of a fluid element with a density that differs from the surrounding fluid by $\Delta\rho$. The temperature gradient of the density is $b = d\rho/dT$.

Index

W

Y

About the Author

Ed Bair is a 3rd generation Coloradoan born in 1922. Summers spent with prospectors on his uncle's gold mining properties formed a basis for independent minded iconoclasm. The area is a museum of geologic history from uplifted mountain peaks that expose eons of sedimentary layers to granite batholiths swept clean by ice flows. A contract to cut timber for a water diversion project in a rain forest at 10,500 feet financed his first year of college.

Chemistry at Colorado State University led to jobs that financed further education. They included analyzing limestone at a quarry in Wyoming, analyzing cattle blood for a university research project at a Wyoming size ranch where the vehicles navigated by compass, analyzing coke coal to make lime from limestone, and analyzing sugar refining at all stages from incoming beets to the pH that allows sugar to solidify as crystals instead of a glass. On enrolling for advanced ROTC in his junior year the Dean withheld his signature pending a "choice" between science and field artillery.

He arrived at the Manhattan Project in Oak Ridge with a fresh B.S. in chemistry in time to help set up facilities that purified the 235U for the first atomic bomb. After World War II he received a Ph.D. from Brown University in 1949 under Professor Charles A. Kraus for a study of incipient micelle formation using electrolytic conductivity. He then accompanied Professor Paul C. Cross to the University of Washington for a post doctoral appointment to set up a molecular spectroscopy laboratory.

His independent academic career began at Indiana University in 1954 where he has remained except for brief excursions as a visiting scientist at the places such as the National Research Council, Ottawa, Cambridge University, England, and the Lawrence Livermore Laboratory. He built a laboratory to study fast processes and energy distributions in the photochemistry of molecules such as ozone. These are the subject of much of his published work. He is now an emeritus professor.

Visits to most of the solar energy facilities in the U.S. as a solar power consultant led to his interest in the future of electric power and a realization that the magnitude of the problem is vastly underestimated.